EFFECTIVE TECHNICAL COMMUNICATIONS

Second Edition

Dan McAuliff

Claire O. Caskey

A. Wayne Bennett, Ph.D.

Vera B. Anand

William E. West

Ernest G. Baxa, Ph.D.

J. Campbell Martin

Robert D. Holstead, Ph.D.

CLEMSON UNIVERSITY

 GINN PRESS

Cover Design by Geneviève T. Parent

Printed in the United States of America.

10 9 8 7 6 5 4 3 2 1

ISBN 0-536-57614-9
BA 4845

GINN PRESS
160 Gould Street/Needham Heights, MA 02194
Simon & Schuster Higher Education Publishing Group

73135

ACKNOWLEDGEMENTS

§ 8.19

The support and encouragement of Mr. and Mrs. R. S. Campbell in the development of the Effective Technical Communications Program and this manual are gratefully acknowledged. The authors would also like to thank Dr. Charles Jennett, Dean of Engineering, and Dr. Robert Waller, Dean of Liberal Arts, for their support.

Program Concept: A. Wayne Bennett

Editorial Supervision: Dan McAuliff

Section	Primary Contribution by
Preface and Section A	A. W. Bennett
Section B	C. O. Caskey
Section B	E. G. Baxa
Section B	J. C. Martin
Section B	R. D. Holstead
Section C	D. McAuliff
Section D	V. B. Anand
Section D	W. E. West

TABLE OF CONTENTS

Page

(Continued)

TABLE OF CONTENTS

(Continued)

This book belongs to _____

_____

_____   Phone _____

PREFACE

An Important Message to Students

Most of your time as an engineering student is devoted to acquiring the technical expertise you will need to become a successful engineer. However, you should keep in mind that the most frequent criticism of young engineers by their managers is their inability to communicate technical information. Outstanding technical advances are sometimes overlooked because of someone's failure to communicate the details effectively. Gaining the approval and support for your project, and your success in the company, will depend largely upon how effectively you communicate your ideas.

Engineering students are often heard to say, "I don't want to get involved with sales". Be aware that you are "selling" yourself when you interview for a job, selling your design when you present it for review, and selling your company when you talk to a customer. Good communication skills are essential to your success.

The Effective Technical Communications (ETC) program has been designed to help you develop good written, oral, and graphic communication skills. Intended for self-study, the ETC manual includes examples that will help you in college and later in your work as an engineer.

Because the ETC program is integrated throughout the engineering curriculum, you will not work your way through this manual section by section. Instead, the assignments will come from parts of the manual most appropriate to the course in which you are given a communication assignment. However, since Section A is an introduction to the manual, read it before using the remaining sections.

There is a tendency to associate information from a particular course with that course only. The material presented in this manual, however, is integral to almost everything you do. Consequently, you are encouraged to use this manual as part of your daily routine. The methods of good technical communication apply to every communication situation.

This manual will present communication skills clearly, and you will use it often as a reference book throughout your college and professional careers. Your technical contributions will be greatly increased by effective communication.

An Important Message To Instructors

The most common complaint about engineering graduates has been their lack of communication skills. Many companies have had to hold special seminars or establish internal programs to improve the oral, written, and graphic communication skills of their employees. This problem has also received a great deal of attention at colleges and universities.

This manual was written as part of the Effective Technical Communication (ETC) program at Clemson University and represents a unique approach to improving the communication skills of engineering students. Since a successful program depends upon the effectiveness of the faculty and staff, it is very important that all instructors understand the ETC philosophy and its relation to the engineering curriculum.

The Effective Technical Communication (ETC) program is an integrated approach for improving the written, oral, and graphic communication skills of engineering students. It is designed to be layered or distributed throughout the undergraduate curriculum. It is intended to supplement rather than to replace courses on technical writing, public speaking, and graphics, the most common vehicles for improving communication skills.

Depending entirely on traditional communication courses has disadvantages. Students have limited technical material to communicate early in an undergraduate program but miss many opportunities to practice if communication courses are scheduled late in the curriculum. The most effective way to improve the communication skills of engineering students is to require the use of good communication practices throughout their college experiences.

Because engineering curricula are already overflowing, it is unreasonable to expect that significant amounts of time can be taken from existing faculty and courses. If an integrated approach is going to be effective, means for efficient delivery of the material must be provided along with assistance in evaluating the work of the students. The ETC program was developed with these concerns in mind.

The nucleus of the ETC program is this manual, which utilizes self-paced instruction and provides examples of effective communication concepts. Evaluation of both oral and written presentations will be done by references to "error messages" listed in Chapters B8 and C5. Use of these shorthand messages should help teachers provide standardized feedback to the student. Written assignments will be judged twice - once by the teacher for contents and again by a writing specialist. Thus, the practice of communication skills can be incorporated into almost all engineering courses by selecting appropriate sections of the manual for assigned exercises.

As an example an instructor would announce at the beginning of a term that a technical report on some aspect of the course must be submitted by midterm, and class time would be avoided by referring students to appropriate sections of the ETC manual. The instructor would also inform the class if he prefers a format other than the manual's format, since forced uniformity throughout the curriculum is not intended.

The ETC manual should also be helpful in engineering practice after graduation. In fact, many of the example letters were paraphrased from letters written by the authors during their careers. The self-paced style should facilitate the use of the manual as a handbook.

As noted earlier, the effectiveness of the ETC program and this manual is determined by you, the instructor. Producing engineering graduates with good communication skills is important, and the program deserves our best efforts. You are encouraged to read and to use the material in this manual. Your comments and suggestions for improving the program would be appreciated. It is worth noting that all of us who teach and work with students should follow the manual's concepts and "practice what we preach." Effective Technical Communication is important for engineers, and now is the time to begin.

Summary Of Contents

The first section of the ETC manual, identified by the letter "A", provides a step by step method of analyzing communication opportunities along with a specific example. Reference is made to this basic analysis throughout the manual.

The second section, identified by the letter "B", is devoted to written communications. This section presents the application of the analysis process to written communications, along with guidelines and examples for the following communication methods: memos, notes, letters, proposals, technical reports, and resumes. The section also includes a brief summary of grammar and punctuation and introduces four character error messages to identify specific errors or problems and suggest corrective actions.

The third section is devoted to oral communications and is identified by the letter "C". The analysis procedure is applied to oral communications. Guidelines and examples are included for the following situations: conversation, telephone use, conducting meetings, and technical presentations. The section includes a section on error messages for oral communication.

The final section, identified by the letter "D" covers graphic communications and utilizes the analysis procedure presented in Section A. Guidelines and examples are included for the following situations: lettering, sketching, scale drawings, tables, charts, and graphs (including computer graphics). The section includes engineering applications of photography, photocopiers, and slide/audio tape presentations. Error messages applicable to graphic communications are included in Sections B and C.

Error Codes and Grammar Guide

In addition to guidelines and examples for improving your communication skills, this manual contains a listing of error codes for written, oral, and graphic material. The error codes will be used by faculty and staff to provide feedback on your work, and you should look up each code in the appropriate section and take the suggested corrective action. Also use the error codes to determine the more important aspects of communications and to critique fellow classmates on joint projects.

The manual also contains a brief summary of the more common mistakes in grammar and punctuation. Mistakes in spelling, grammar, and punctuation will give the impression of an uninformed and careless person and undermine the audience's confidence in your work. Think about it; would you put your money in a bank that makes mistakes with your statements? The memos and letters you write, the presentation you give, and the drawings you send out carry a message about you. Make sure it is what you intend.

Section A
ANALYSIS OF COMMUNICATION OPPORTUNITIES

The most important skill of a successful engineer is the ability to communicate effectively. Many practicing engineers do not utilize higher level mathematics, many do not utilize thermodynamics, many do not utilize circuit analysis, but every successful engineer is an effective communicator.

A1: STEP-BY-STEP ANALYSIS

You should begin and end every communication opportunity with careful analysis of the participants and the situation. Initially, it should be a very deliberate, detailed process. As you gain experience, many of the details can be omitted, but you should analyze every situation. It is the key to effective communication. The four-step analysis process is to determine your objectives, analyze your audience, make an in-process analysis, and make a post-communication analysis.

A1a: Step 1: Determine Your Objectives

The first step of the analysis is to determine your own goals or objectives. What are you trying to accomplish? If you have several objectives, rank them. For example, in presenting an oral design review for your senior project, your goals may be the following:

(1) get good grades,

(2) sharpen your presentation skills, and

(3) recognize the contributions of the other students who worked in your group.

The failure to concentrate on goal (1) might lead you to leave out important material or inadequately prepare.

If you don't make a deliberate effort to sharpen your presentation skills, goal (2), you may not try out a new technique that should be developed before you go into industry.

Failing to recognize the contributions of your fellow workers, goal (3), can lead to hurt feelings and is typical of the things that are often forgotten if left to the last minute. You should write down all of your goals and check them off as you work them into your presentation.

A prioritized listing of your goals or objectives is equally important for written and graphic communication. In these cases, the priority of a given goal may determine where it appears (and how often it is repeated) in a written document, or how large, what color, and where it appears in a drawing, chart or photo.

A1b: Step 2: Analyze Your Audience

Analyzing the audience (listener, reader, or viewer) involves several steps, beginning with the determination of their goals and objectives. Why are they going to listen to your presentation, read what you are going to write, or look at the drawings, charts, or graphs you are preparing?

In addition to knowing the general goal(s) of your audience, you must know the degree of their interest. Knowing when and where to present detailed coverage requires a careful analysis of your audience. You don't want to bore someone with detail in which they are not interested or fail to provide information they desperately need or want.

The analysis of your audience should also help you determine the vocabulary and technical level of the material you are preparing. Using highly technical terms and vocabulary in writing or speaking, or including complex scientific symbols on drawings and charts, will be lost on a non-technical audience and may have a negative impact. Suppose you have just spent 6 months designing a new system and know every detail about its construction and operation. You must not point out and dwell on those fine technical features when your audience is only interested in the general operation characteristics of the system. Even though you know better, this is one of the most common mistakes made when presenting technical material.

By now it should be obvious that the more you know about your audience, the better you can communicate. Ideally, you should know the "personality" of your audience, their educational backgrounds and career fields, and maintain a file on an audience with whom you have regular contact.

A1c: Step 3: In-Process Analysis

The determination of your goal and those of your audience should be a continuing process since they are subject to change. Therefore, you should be aware of the impact of your first letter, the initial oral presentation, or the first series of drawings and make appropriate changes in subsequent communication.

An advantage of face-to-face communication, or oral presentation, is the opportunity to judge reactions and make immediate changes in your level of coverage, vocabulary, or emphasis. This also indicates that considerable care is required for written documents or graphic materials that will be read or viewed without the benefit of your explanation. Take the time to review the material from the point of view of the person or group with whom you intend to communicate.

A1d: Step 4: Post-Communication Analysis

One of the most effective ways to improve your communication skills is to analyze the results of each interaction after it is completed. You should go back to the original list of your objectives and the audience's objectives and evaluate how well you accomplished them. Some of the questions you should ask are as follows: Why did you do poorly? What went well? How can you improve? How effective was in-process evaluation? Track yourself as you improve, because effective technical communication is fundamental to a successful career.

A2: AN EXAMPLE SITUATION

Assume that you are employed in an engineering organization involved in developing a new test chamber for testing new materials. The system will utilize a computer to control and monitor temperature, humidity, air flow, and light in a 2 ft. x 2 ft. x 2 ft. chamber. One of the first communication opportunities would be to schedule laboratory space for construction of a prototype. This will require that you write a memo to communicate your needs to the laboratory supervisor, and your analysis might proceed as follows:

A2a: Step One: Determine Your Objectives
1. Reserve an appropriate amount of space in the laboratory along with the needed electrical and mechanical support services.
2. Arrange for the anticipated laboratory technician assistance.
3. Communicate the importance and priority of the project.

A2b: Step Two: Analyze the Audience
Objectives of your audience (the laboratory supervisor)
1. Schedule laboratory facilities to meet the needs of the company as determined by senior management.
2. Assure safe operation of all laboratory facilities.
3. Provide the support deemed necessary by the engineering staff.

Audience Profile
1. The laboratory supervisor is a very busy engineering person who understands the construction and testing of systems but has very little design experience.
2. He is fully aware of the hazards of electro-mechanical testing.
3. He is intolerant of young, inexperienced engineers.
4. He is very conscious of maintaining schedule.
5. He is very proud of the laboratory safety record.

You now have enough information to assemble the material for communicating your needs. Since this situation should be documented carefully, a written memo (see Chapter B1) or letter (see Chapter B2) should be used, and some drawings may be required (see Section D).

Your written document should state the relative priority as determined by senior management and explain any safety hazards. Clearly identify the starting and ending dates for the testing, along with necessary equipment.

Specify the electrical, heating, and ventilating requirements and include drawings of special fittings and connections.

In view of the laboratory supervisor's awareness of safety, include a few "extra" comments regarding hazards and assurance of meeting the schedule stated. Compliment the fine safety record the laboratory has maintained and avoid turning him off by over-emphasis on details.

A2c: Step 3: In-Process Evaluation

In this instance, your memo or letter should be followed by a phone conversation, or second memo, to confirm the arrangements. Check to see if all of your objectives have been achieved and be aware of the attitude of the laboratory supervisor to see if you established a good working relationship. Perhaps additional interaction is needed.

A2d: Step 4: Post-Communication Analysis

After the testing is complete, determine if everything went according to schedule. Ask yourself the following questions: Were all requirements properly identified? Was additional information needed? Were your drawings correct and adequate for the task? How could you have made the testing run more smoothly? Summarize your effectiveness as a communicator for use on the next project. Learn from your good and bad results.

A3: REMEMBER

All engineers will need to communicate effectively throughout their careers, and you should use the teaching presented in this manual in everything you do. Keep the manual as a desk book for the remainder of your college career and when you go to work. Your technical contributions will be measured by how well you communicate.

SECTION B
WRITTEN COMMUNICATIONS FOR THE ENGINEER

As an engineering student you will have to write laboratory reports, letters, memorandums, proposals, progress reports, and formal reports on projects undertaken. You need to be able to write <u>clearly</u> and <u>concisely</u> to convey information to your teachers and fellow students. Your work may be exceptional, yet it will do you little good if you cannot explain it to others.

When you finish school and accept a job, writing will take on an added importance. A college degree will help you get a job, but your advancement will depend heavily upon your ability to write well. Engineers spend as much as half their working time writing, and often their written reports are their only contact with superiors.

This section begins with a discussion of memos and notes and how they should be used. This is followed with explanations and samples of a wide variety of business letters, reports, and resumes. The section concludes with chapters on punctuation, "error messages," and references for further study.

CHAPTER B1

MEMOS AND NOTES

B1a: ANALYZE FOR BETTER COMMUNICATIONS

As an engineer, you must constantly communicate with others in your organization to keep them properly informed. This type of interaction is referred to as internal communications, and the principal vehicles will be MEMOS and NOTES. As in every communication situation, you should review and apply the analysis procedures presented in Section A. The importance of determining your objectives and analyzing your reader(s) cannot be over emphasized. A check list for this topic is as follows:

(1) Determine your objectives.

(2) Determine the objectives of your reader(s).

(3) Select the proper format.

(4) Choose the appropriate vocabulary.

(5) Set the proper tone.

(6) Evaluate your results.

B1b: THE MEMO

Somewhat less formal than the business letter, which is covered in Chapter B2, the memo is a semiformal communication between members of the same organization. It may be a single paragraph or several pages, and most organizations have printed memo forms containing designated lines for "DATE, TO, FROM, and SUBJECT."

B1c: When to Use a Memo

You may use the memorandum for any type of internal communication, but it is semi-formal and may not be as suitable as a NOTE in certain situations. However, the memo, rather than the note, should definitely be used in the following situations:

• A written file record is required.

• The addressee is a department or group.

• Carbon copies are designated for other persons or departments.

- The addressee is more than one organizational level removed from you.

B1d: How to Write a Memo

Remember the following things when you write a memo:
- Plan it with purpose and reader in mind.
- Be sure your subject line announces the subject clearly.
- Be as brief as possible consistent with your purpose.
- Adjust the tone to your reader by writing more formally to those above you in the organization.
- Use headings and subheadings to call attention to your main topics and to change topics.
- Use lists when possible to call attention to important points.
- Include an introduction to provide background information if your reader is not familiar with your subject.
- If practical, attach copies of any documents to which you refer.
- Place your initials beside your name.

B1e: THE NOTE

The NOTE is a brief unstructured communication between two persons who are usually within the same organization.

B1f: When to Use a Note

You should use the NOTE for most internal one-to-one written communications that do not require file documentation. It is usually, but not necessarily, handwritten and should always be brief (less than 50 words). A computer network permits individuals to exchange electronic notes.

Memos and business letters can also be handwritten, but they are not notes.

Since informality of the note makes it level sensitive within an organization, confine your use to peers and to those no more than one level above or below. Do not send notes above your immediate superior or below your immediate subordinates unless persons are known on a personal basis.

<u>B1g</u>: <u>How to Write a Note</u>

A note can be on a separate sheet, usually smaller than 8 1/2 x 11, or it can be in the margin on a copy of the document being discussed. In either case, write or print the message clearly and, if on another document, mark those areas to which the various notes apply.

Warning: The note is informal, but do not assume that the receiver is a mind reader. Not everyone recognizes your writing. Unless you are using a personalized note pad, <u>sign</u> and <u>date</u> notes just like any other communication.

<u>B1h</u>: <u>Pre-printed Notes</u>

There are a variety of pre-printed notes on the market for items like telephone messages or repetitive instructions, such as "FILE" or "Follow-up." These can be very time saving, so do not hesitate to have your own special messages pre-printed for such repetitive tasks.

<u>B1i</u>: <u>Self-reply Notes</u>

These notes usually have a carbon copy, and the receiver is to write an answer and send a copy back to you.

This type of note is excellent for situations where they are used as standard operating procedure, but some people find the "hurry up and answer" implication too offensive for the occasional note. Use such self-reply notes with discretion.

<u>B1j</u>: <u>REMEMBER</u>

Analyze the situation and determine whether you need a <u>memo</u> or a <u>note</u>. A memo used in a situation calling for a note will seem stuffy, and the opposite situation will seem impertinent, or at best, unprofessional.

CHAPTER B2
BUSINESS LETTERS

B2a: <u>ANALYZE FOR BETTER COMMUNICATIONS</u>

Writing effective letters will play a big role in your career as a technical person. These may range in length from one short paragraph to several pages and range in complexity from a simple letter of transmittal to a complicated explanation of sophisticated equipment. Before writing, analyze the opportunity as outlined in Section A. Listed below are some of the more common types of business letters:

Letter of transmittal

Letter of application

Letter of complaint

Letter answering a complaint

Letter of refusal

Letter of acceptance

Letter of inquiry

Letter responding to inquiry

Letter of instruction

Letter to customer--following phone contact

B2b: <u>LETTER OF TRANSMITTAL</u>

When you send a report to someone, it is good to have a letter of transmittal with it to <u>convey</u> the report from the writer to the reader. This letter is often clipped to the cover of the report; at times it is sent separately. In many ways it serves the same purpose as a preface in a book. Following are some of the things that can be included:

1. Give the title of the report on a subject line or in the first paragraph.

2. In the first paragraph refer to the occasion of the report--a progress report, for example--and tell why the report is being submitted. You may also refer to a request, a contract, an informal agreement, or whatever is appropriate.

3. In the second paragraph explain the purpose and scope of the report and make other appropriate comments. Additions or deletions, for example, might need to be mentioned.

4. Acknowledge any help you have received in doing the work you are reporting.

5. In closing you can express your appreciation for the opportunity to do the work and the report. You can also offer to answer questions that might arise. Some writers end with the <u>hope</u> that the report will be satisfactory. It is better to be <u>sure</u> it's satisfactory and give no implied suggestion that it could be otherwise.

March 10, 1985

Mr. James McDougal, President
Thermal Engineering
Box 9021
New Orleans, LA 632014

Dear Mr. McDougal:

Here is our report entitled <u>The Feasibility of Building a Second Ball Bearing Plant in Easley, S.C.</u>, which we promised you by March 15, 1985.

| What is being transmitted.

We added a section on the labor union situation in the area, feeling that it might be of interest to you. The other sections on state and local taxes, area unemployment, utility costs, local wage scales, and available technical support have been up-dated.

| Purpose and scope of document being trans-mitted.

The Easley Chamber of Commerce was most cooperative in helping us gather data for this report.

| Acknowledge help.

Many thanks for letting us do this study for you. If we have raised questions for which we provided no answers, please let us know.

| Express apprecia-tion and offer to answer questions.

Sincerely,

B2d: LETTER OF APPLICATION

When applying for a job you will need to write a letter of application accompanied by a resume, Chapter B6. Make the letter brief, no more than a single page, and be sure it is typed correctly, including proper spacing of parts. Address the letter to a person, not "Gentlemen" or "Dear Sir". Use a colon after the salutation. The opening paragraph should capture the reader's attention by referring to an advertisement, a person from the company who told you about an opening, one of your professors who suggested that you write, or whatever stimulated you to apply. It could be to a company with no known openings.

In the second part of your letter, interpret, expand, or comment on items from your resume that apply to the job you want. Show, for example, how particular work experiences have prepared you for the job. Or discuss courses which may make your qualifications unique: computer courses, foreign languages, mathematics beyond that required, and courses that are broadening such as philosophy, psychology, history. This is the place, also, to emphasize your abilities, inclinations, aptitudes. Mention your grade point ratio if it is especially high. This is the sales section of the letter. Try to make the reader feel that you are someone worth talking with in an interview. The ultimate purpose of the letter and resume is to get an interview.

Finally, make a specific bid for an interview, rather than asking the person to call you if interested. Mention your free afternoons or spring holidays as good times, but leave it open since you will want to go regardless of convenience to you.

February 6, 1985

Mr. James Benson
Personnel Director   Address to a
McCrory Robotics   person.
Denver, West Virginia

Dear Mr. Benson:

    Dr. Benton Dawkins, professor of mechanical
engineering at Clemson University, suggested that
I write you concerning a job with McCrory. He
said that at a recent meeting in Atlanta you   Capture
asked him to recommend graduating mechanical   attention.
engineers who could benefit your company.
Graduating May 15, 1985, I am one of those he named.

    For as long as I can remember I have been
interested in things mechanical---how they
work and how they could be made better. Here
at Clemson I have done well in all of my   Interpret
courses, but particularly in design as you   and expand
can see from the enclosed transcript. In   resume.
addition to the regular curriculum for
mechanical engineers, I have taken electives
in robot design and in computer engineering.

    Having lived in Japan for three years, I
learned to speak Japanese fairly well, and working
at Kusoi Robotics in Sendai for three months
helped my understanding of robots as well as the
language. Additionally, I spent last summer
working with a robotics team at Platt Saco
Lowell in Easley, S.C. We used some of your
equipment in our experiments.

    I am quite interested in the possibility of
working for your company and would like very much
to talk with you about my qualifications. My   Make specific
spring holidays are March 11-15. If any time   bid for an
during this week would be convenient for you,   interview.
please call me. If some other time would be
better for you, I will come then.

                Sincerely,

B2f: LETTER OF COMPLAINT

As an engineer, you may have to write or assist your Purchasing Department in writing a technical Letter of Complaint.

There are several key elements in such a letter:

1. Give proper reference to Purchase Order and Sales Order Numbers.

2. Identify the specific product or service occasion and associated dates.

3. State the technical failure as <u>you</u> see it.

4. Request directions on what you should do next.

5. If appropriate, ask what will be done to reduce future complaints of this same nature.

6. Keep your Purchasing Department informed.

B2g: <u>SAMPLE LETTER OF COMPLAINT</u>

NORTH TEXAS POWER COMPANY
PO BOX 4567
GRAND PRAIRIE, TEXAS 75124

June 10, 1985

Continental Instrument Company
Box 1234
Dallas, Texas 75421

ATTN: Mr. E. F. Jones
        District Manager

REF: Our P.O.-85-5678; Your S.O.-12345678 | Proper reference to order Numbers

Dear Mr. Jones:

We received the above shipment of 4 TYPE B6TS Polyphase meters, Catalog No. 36479, on May 31 and find one of these instruments to be inoperative. | What product and when received.

It appears to have an open circuit on the rear potential coil. | Specific nature of failure.

If you wish us to return this unit to your factory, please provide a Return Authorization Number. Otherwise, we will hold in the Meter Dept. for your inspection. | Tell us what to do.

Since this is the third such meter received this year, we would appreciate a report from your Quality Assurance Dept. on factory steps being taken to reduce such occurrences. | What are you doing to keep this from happening again?

Very truly yours,

John Doe
Engineering Manager
Meter Dept.

cc: Mr. S. F. Smith
Purchasing Dept. | Always keep your own Purchasing Department informed.

B2h: ANSWERING A COMPLAINT

Your Sales Department will often need to answer technical complaints. Such letters are sometimes written by Sales Engineers and other times by Quality Engineers or other manufacturing personnel.

There are several key elements required in such a letter:

1. Give proper reference to Purchase Order and Sales Order Numbers.

2. Thank the complaint writer.

3. Tell him his immediate alternatives.

4. Do not initially admit or deny responsibility or negligence unless it is obvious.

5. Offer to "go out of your way" to ease his immediate problem.

6. Assure him that all questions will be answered.

7. Close on a friendly note.

B2i: <u>SAMPLE ANSWER TO A COMPLAINT</u>

CONTINENTAL INSTRUMENT COMPANY
TEXAS DISTRICT
BOX 1234
DALLAS, TEXAS 75421

June 16, 1985

North Texas Power Company
PO Box 4567
Grand Prairie, Texas 75124

ATTN: Mr. John Doe
       Engineering Mgr.
       Meter Dept.

REF: Your Letter of June 10, 1985
      PO-85-5678, SO-12345678

Dear Mr. Doe:

Thank you for letting us know about the above meter shipment.

The defective meter should be returned to our factory as soon as possible so that our Quality Dept. can determine the exact nature of the problem.

You may return the meter C.O.D., using the enclosed Return Label on the carton, or hold for my regular Grand Prairie visit next month.

I'm sorry that we do not have a replacement meter of this same Catalog Number in Dallas stock, but I can have one transferred from San Francisco within two weeks if our normal 30 day factory turn-around is not satisfactory.

I apologize for the defect and will request a full report from the Quality Manager as soon as the returned meter has been examined.

Timely reports such as yours allow us to maintain the best possible quality.

Very truly yours,

E. F. Jones
District Mgr.

Proper reference to complaint letter and Order Numbers.

Thank him for writing.

Tell him alternatives on what to do now. Don't admit the defect is your fault, as yet.

Offer to go out of your way if necessary.

Assure him that his question on future occurrences will be answered. Close on a friendly note.

29

B2j: LETTER OF REFUSAL

During your professional career, you will often find it necessary to communicate your refusal of a position with another company, a new job assignment with your current employer, or committee membership or office in a technical society. You may also need to decline lending your equipment, facilities, or personnel. It is important that you carefully consider each opportunity and decide whether or not you wish to accept the offer. Once you have decided that it is not in your best interest to accept, your decision should be communicated promptly. The manner in which you refuse often affects your chances for future opportunities. Your letter of refusal, therefore, should be carefully constructed.

Section A of this manual emphasized the importance of the analysis of each communication opportunity. In applying the techniques covered, your analysis of this communication opportunity should include the following:

1. What are your objectives in refusing the offer?
2. What were the objectives of the person or organization making the offer?
3. Are there intermediate objectives that your letter should mention as a compromise?
4. How can your refusal of the offer be best communicated to benefit everyone involved?
5. Make sure that the appropriate vocabulary is used and that the tone of the letter conveys the message you intend.

The key elements that your letter should include are as follows:

1. Indicate that you considered the offer carefully.
2. Make your decision known near the beginning of the letter-don't leave your reader "hanging."
3. Provide one or more good reasons why you are refusing.
4. Make it known that you would like to be considered for other opportunities, if that is the case.
5. Express appreciation for the reader's consideration and offer to be of assistance if appropriate.

B2k: <u>SAMPLE LETTER OF REFUSAL</u>

806 Ridgecrest Dr.
Clemson, SC 29631
May 12, 1985

Mr. David Tice
Director of Engineering Personnel
Technologies, Inc.
P.O. Box 324
Beaumont, TX 65382

Dear Mr. Tice:

Thank you for your letter of May 5, 1985, offering me a position with your firm. I have carefully considered this opportunity and find it very appealing. However, I am not in a position to accept your offer at this time.

Indicate careful consideration.

State your decision.

The decision not to accept your offer was very difficult. I am aware that your company is heavily involved in high-tech projects and will be expanding rapidly in the future. I also recognize the importance of the opportunity to learn a new programming language. Since I have been in my present position for only six months, I feel that it would be unwise for me to change jobs at this time.

Indicate interest.

Give reason for not accepting.

I am very impressed with your company and enjoyed meeting you and your associates. I wish you success with the new project and hope that you will keep me in mind for future opportunities. Thank you for considering me.

Express interest in future jobs.

Show appreciation.

Sincerely,

George Richardson

<u>B21</u>:   <u>LETTER OF ACCEPTANCE</u>

During your engineering career, you will be offered opportunities that you may wish to accept. In many instances, the level of involvement is minimal, and a simple yes or no is all that is necessary to convey your decision. There are, however, many instances in which a written response is required. Also, there are situations in which a letter of acceptance is needed to document your level of involvement, specify conditions under which you accept the offer, or state limits on your responsibility. All too often, your eagerness to get involved will cause you to accept an offer without taking the time to write an appropriate response letter. Keep in mind that how you accept an offer can be an important factor in your success in carrying out the tasks involved.

Section A of this manual emphasized the importance of the analysis of each communication opportunity. In applying the techniques covered, your analysis of this communication opportunity should include the following:

1. What are your objectives in accepting the offer?

2. What were the objectives of the person or organization making the offer?

3. Are there special conditions that your letter should mention?

4. Make sure that the appropriate vocabulary is used and that the tone of the letter is very positive.

The key elements of a letter of acceptance are as follows:

1. State your acceptance at the beginning of the letter.

2. Indicate enthusiasm for whatever it is you are accepting.

3. Specify the limits of your responsibility.

4. State any special conditions on your acceptance.

5. State the outcome if appropriate.

6. Express appreciation for the opportunity.

B2m: SAMPLE LETTER OF ACCEPTANCE

HYBRID ELECTRO-SYSTEMS, INC.
1 TECHNOLOGY PARK
EATON, OHIO   57334

Interface Design
Room 100E
Building H-5
June 2, 1985

Mr. A. C. Conrad, Manager
Systems Engineering
Room 208, Mail Stop C23
Building E-1

Dear Mr. Conrad:

I enjoyed our phqne conservation yesterday
and was very pleased with your offer to have me
join the Systems Engineering group. I discussed
the situation with my current supervisor, Milton
Lewis, and he has agreed to the change. I feel
that this is an excellent opportunity for me and
am confident that I can provide the computing
expertise needed by your group.

> Indicate pleasure
> with opportunity.
>
> Confirm necessary
> approvals.
>
> Express
> confidence.

Since our discussion covered a number of
topics, I would like to confirm the more
important items. It is my understanding that
my salary will increase to $30,000 per year
and the date of my annual review will continue
to be April 1. I also understand that I will
be responsible for the simulation software and
will have two computer programmers working
under my supervision.

> Confirm important
> conditions of
> your acceptance.

I am excited about working with your
group and look forward to moving into the
Systems Engineering Office complex on
July 1, 1985. Please let me know if
there is any thing I need to do in the
meantime. I appreciate you interest in
me.

> Reaffirm pleasure
> of opportunity.
>
> Confirm important
> dates.

Sincerely,

Gilbert Hayse

cc: Mr. Milton Lewis

> Copy to others.

B2n: LETTER OF INQUIRY

It will often be necessary for you to request information in writing so that a complete record of inquiry/response is obtained. However, you should always carefully weigh the need to handle an inquiry by letter as opposed to other less expensive and perhaps more timely means. If a letter seems appropriate, it is important that you provide all of the information needed by the reader to respond intelligently. The tone of the inquiry should be courteous, well-mannered, and create a pleasant atmosphere since letters of this type frequently lead to further business contracts. As a guide for clarity it is good practice for you as the writer to put yourself in the reader's place to see if your letter is complete and clear. The following key elements should be included in the letter of inquiry:

1. State carefully the circumstances which necessitate the inquiry.

2. State the facts needed by the reader in making a complete reply.

3. Ask for the information or state the questions.

4. Express gratitude for the favor asked.

B2o: SAMPLE LETTER OF INQUIRY

NORTH MANUFACTURING COMPANY
"industry leaders"
2000 Second Street
Middletown, South Carolina 34567-1234
808-221-9000

May 15, 1984

Mr. I. William McKall
Field Services Manager
American Computer Company
Silacon Vallee, CA 22264-4590

Ref: Purchase Order PO 2315-85 dtd 2 Feb 1984

Dear Mr. McKall:

We recently purchased twenty personal
computers model PCOMP-II which include 512k
RAM, two floppy disk drives, the high
resolution color video monitor, and the
model 22 near-letter-quality dot matrix
printer. At the time of purchase no
maintenance agreement was executed.

Mr. Ted Billus of your sales department
told us during purchase negotiations that
you have some limited maintenance contract
support available and that arrangements can be
made any time during the original six month
warranty period. We have decided that we
want contract maintenance for these machines
on a one-year renewable basis.

Can you provide such a maintenance plan and
will you give us a quotation? The plan should
provide maintenance for all twenty machines and
associated peripheral equipment. Please include
your projected maintenance call response time
as a part of your quotation.

If you are unable to provide contract maintenance for us, I will appreciate any information that you can give me concerning
possible field service support from other
vendors.

If you need additional information please
call me (808-221-9000 ext. 235). I will
appreciate your prompt attention to this
matter. Thank you.

Sincerely yours,

J. P. Roscoe, Manager
Computer Support Design

Annotation
Reference to purchase.
Facts needed for response
Circumstances which necessitate inquiry
State questions and expectations.
Additional expectation in response.
Gratitude for request response and offer added help.

B2p: RESPONSE TO LETTER OF INQUIRY

In responding to a letter of inquiry you should try to adapt in style and language to the inquiry letter so that the reply will be understood. This will normally require your personal attention in drafting the letter as opposed to using a form letter response. You should be prompt, systematic, and above all be <u>responsive</u>. If response material is not immediately available, it is always good practice for you to promptly acknowledge the inquiry and set a date for a final reply. As with the letter of inquiry, your response should be courteous and well-mannered because correspondence of this type frequently leads to further business contracts. The following key elements should be in the response to a letter of inquiry:

1. Acknowledge the inquiry or state the circumstances necessitating the reply.

2. Answer the questions fully.

3. Build goodwill and pave the way for further contacts.

B2q: SAMPLE RESPONSE TO INQUIRY

AMERICAN COMPUTER COMPANY
SILACON VALLEE, CA 22264-4590

FIELD SERVICES DIVISION
909-403-8700

May 29, 1984

Mr. J. P. Roscoe
Computer Support Division
North Manufacturing Company
Middletown, SC 34567-1234

Ref: yr ltr dtd 15 May 1984

Dear Mr. Roscoe:

In response to your letter concerning a
maintenance contract agreement for the PCOMP-II
personal computer systems, we no longer offer
contract maintenance services. As of April 1 we
provide maintenance only on a per-call basis.
We believe that we can provide satisfactory
response time and considerably reduce your
maintenance costs by operating on a per-call basis.

> Identify
> inquiry.
> State
> response.

We have provided additional staff at each
of our field service centers to ensure rapid
response to maintenance calls. Our average
response time has been less than one working
day. Service to you can be provided at a
$38.00 hourly rate based upon time of
departure and time of return, plus parts, and
plus travel expense from our nearest service center
in Columbia. A more detailed description of
our rates is included in the attached brochure.
I will be happy to answer any questions that
you may have concerning our new maintenance plan.

> Expand on
> response and
> since it
> may not be
> readily
> acceptable
> qualify and
> sell it.

If you still prefer a contract maintenance
plan, Midus Computer Shops will provide this
service for owners of American Computer Company
products. Information may be obtained by
calling either your nearest Midus Computer
Shop or by contacting their headquarters at
P.O. Box 2345, Boston, MA 34567-8110.

> Offer alter-
> native to
> satisfy ex-
> pectation of
> inquiry and
> to promote
> future
> business.

Thank you for your inquiry, and we will
be happy to help you as needed.

> Foster good-
> will for
> future
> business.

Yours very truly,

I. W. McKall, Vice-President
Field Services Division

B2r: LETTER OF INSTRUCTION

As a technical person, you will be asked to instruct others who need your specialized knowledge to perform a particular task. The letter of instruction serves this purpose. If the matter is a complex one, the letter may be more conveniently structured as a report; however, a letter or memo will often suffice. You should provide explanatory background for the letter of instruction to aid the understanding and retention of the reader. Make use of the imperative mood for giving specific instructions and use the imperative verb form in stating commands or strong requests. Be careful that you do not appear unduly brusque or imperious particularly with clients or persons outside your organization. Be sure that your instructions are complete and that your statements are specific and clear. Whenever directions are given you should arrange the presentation in logical steps. Consider using such devices as numbering, underlining, and indention of headings to help clarify the indicated procedures included in a letter of instruction. Key elements include the following:

1. Provide background and reasons for instructions.

2. State specific instructions clearly, concisely, and in logical order.

3. Use imperative mood.

4. State how added or alternative instructions may be obtained.

June 13, 1985

Mr. Oscal R. Wayne
Computer Center
Wesleyan Teachers College
White Plains, OH 34091-3124

Dear Mr. Wayne:

    Thank you for deciding to buy the PP-45
personal computer. A special program,
INSTALL45, is used to install the operating
system. Once this installation is complete for
a particular hardware configuration, it may be
reproduced for operational use in all like con-
figurations. If new peripherals are added later
or other hardware changes are made, the installation program can be rerun as needed. To prepare
the PP-45 system, which includes a 20 Mbyte hard
disk, dot matrix printer, and local area network
(LAN) communications port, the following three
steps should be taken before beginning the
operating system installation procedure.

| Background
| and basis
| for instruc-
| tions.

        1. Connect the printer and the LAN inter-
    face adapter to the console using the
    procedure on page 47 of the "PP-45 Hardware
    Reference Manual." Cables are provided
    in the accessory kit enclosed in the
    console carton.

| Specific in-
| structions
| indented for
| clarity.

        2. Interconnect two such systems using at
    least 10 meters of shielded 22 gauge
    twisted pair wire. Connect the two
    LAN interface adapters as described
    on page C-2 of Appendix C in the
    "PP-45 Hardware Reference Manual."

| Imperative
| mood concise
| precise.

        3. Complete the operating system install-
    ation using the procedure for networks
    which begins on page 16 of the "Operating
    System Installation Manual for the PP-45."

This procedure provides an operating system
environment for networking any number of machines.
My office is available to help if you have any
difficulty with your installation.

| Provide for
| added assis-
| tance.

                Sincerely yours,

                E. S. Weedon, Senior Engineer
                Computer Communication Dept.

B2t: <u>LETTER FOLLOWING PHONE CONTACT</u>

As noted in Chapter C2 of the Section on Oral Communications, the telephone is one of your most important communication tools, and proper telephone procedures are essential to your success. If you have not read Chapter C2, you should do so before continuing with this section. The effective use of the telephone can open many doors and provide exciting opportunities for you, provided <u>you follow up phone contacts properly</u>. This section will cover the key elements of following a phone conversation with a confirming letter.

In developing the material for this section, it is assumed that you have talked with someone by telephone and are writing a letter to confirm the important points of your discussion, and perhaps, introduce additional details for the consideration of your contact. Before you begin writing, take the time to carry out a communcation analysis of your telephone conversation and the letter you are about to write. Section A provides the general guidelines, and you should review that material if needed. The more important elements to consider in your analysis are as follows:

ANALYSIS OF TELEPHONE CONVERSATION

1. Did you achieve all of your objectives?

    a) Did you cover every point of the product or service you were discussing?

    b) Were you able to answer every question raised in the conversation?

2. Did you accomplish the objectives of the individual with whom you spoke?

3. Was the general tone of the conversation satisfactory? Do you need to smooth out a rough spot in your communication?

ANALYSIS OF THE LETTER YOU ARE WRITING

1. Now that you have completed the telephone contact, what are your objectives? For example, they may be to clarify points, raise new issues, or point out additional features.

2. Based on your telephone conversation, what are the objectives of the individual with whom you spoke?

3. Be sure to use the proper terms, maintain an appropriate technical level, and chose words and phrases to set the right tone.

After you have analyzed both the telephone conversation and the communication opportunity of your follow-up letter, you need to assemble the required additional details and begin writing. The key elements of your follow-up letter are as follows:

1. Identify yourself and give a brief summary of the telephone conversation to help your reader recall the details of your call.

2. Indicate your interest and let your reader know that you have pursued questions raised in the telephone conversation.

3. Confirm the important details discussed by telephone.

4. Include appropriate additional information, and avoid clouding the issue with unnecessary details.

5. Offer to provide additional information if required.

6. Set dates for future actions if appropriate.

7. End the letter in a positive (or negative, if appropriate) manner.

SPECIAL NOTE:

In the course of your career, you will need to write a letter following an unpleasant phone conversation. In some instances, you will want to use the letter to correct a misunderstanding. In other cases, the letter needs to confirm the bad news delivered by phone. In all situations, your letter must be clear, direct, and accurate. It is difficult to achieve binding agreement on telephone conversations, but <u>a letter is a permanent document</u>. Be careful!

THERMODYNAMICS, INC.
East Industrial Park
Columbia, SC 29688
803-224-8955

April 19, 1985

Mr. Emmett Strickler, Manager
Engineering Procurement
Satellite Systems, Inc.
P.O. Box 739
Scranton, PA 17355

Dear Mr. Strickler:

I am writing to confirm our phone conversation on April 17, 1985, in which we discussed the new heat shields developed by our company. Our HT-3a units exceed the range you specified and are ideal for the high performance engine you are designing.

Identify date and subject of your phone call.

As I understood your requirements, there are three primary concerns, and this letter will confirm each point. First, the temperature range will be 10°C to 900°C. Second, the the thickness is 1.250" ± .005". Third, the material is available in 6-inch strips that can be cut to length, not exceeding 8-feet.

Confirm important details from call.

After talking with you, I checked with our laboratory technician, and we can supply HT-3a units in black or white. Other colors are being developed.

Provide additional help.

This should provide the information you need, and I will be happy to supply additional details. Based on our conservation, I am confident that there are other projects of mutual interest and look forward to working with you.

Offer additional help.

End in a positive manner.

Sincerely,

Lewis Rich
Materials Engineer

cc: Al Little
     Dick Rankine

Send copies to appropriate people.

42

CHAPTER B3
PROPOSALS

B3a: ANALYZE FOR BETTER COMMUNICATIONS

Your success as an engineer will greatly depend on the proposals you and your associates write. In many instances, they determine whether or not you receive internal funding for a project you want to do. In other situations, they will determine whether or not your company is awarded a contract upon which your job depends. Nothing needs to be analyzed more thoroughly before communicating.

The analysis procedures presented in Section A should be reviewed and applied. The importance of determining your objectives and analyzing your reader(s) cannot be over emphasized. A check list for this section is as follows:

(1) Determine your objectives.

(2) Determine the objectives of your reader(s).

(3) Select the proper format.

(4) Chose the appropriate vocabulary.

(5) Verify every detail.

(6) Carry out a post communications analysis.

B3b: WHAT IS A PROPOSAL?

A proposal is a written offer to solve a problem. Naturally, for engineers it usually means a technical problem. The proposal most often includes how the problem will be solved and what the cost will be. It can be regarded as a selling device, a piece of persuasive writing. However, it relies upon conservative language, a rational approach, and concrete data in the form of charts, graphs, and tables, rather than upon emotional language. Proposals range in length from a few paragraphs to volumes.

B3c: HOW IMPORTANT ARE PROPOSALS?

Ask any group of engineers how many of them have written a proposal recently. If even one says no, it will be unusual. Proposals are a way of life in the business world. Many companies depend upon them for survival, for selling their services. An executive for a construction company may spend most of his time either writing proposals or supervising others who are writing them. A company involved in military systems may spend thousands of manhours putting together a proposal for building the wing for a military airplane. If some other company does a more convincing job, then time and money have been lost.

Internal proposals, those written within a company, may not be as dramatic, but they are nevertheless important. Companies expect their employees to anticipate needed changes--to improve a product, to speed up delivery, to reduce cost. In a typical situation, you study the problem and propose your solution to your manager. Included will be your plan for solving the problem and how much you think it will cost. If your proposal is convincing, it will probably be approved.

As an engineering student you will write proposals for research projects or perhaps a senior design project. You may be asked to serve on a committee with other students and faculty to propose changes in a particular course or in your curriculum.

B3d: WHERE DO YOU START A PROPOSAL?

If your proposal is an unsolicited one, start with the problem itself. Locate it; define it; describe it. Make your reader aware that a problem exists. You might want to tell how the problem arose: from neglect, from circumstances beyond control, from recent changes. Is the company losing money because of it? Have attempts been made to solve it? If you show a thorough knowledge of the problem, then your proposal to solve it will be more effective. Your chance of "selling" your proposal will be better.

If, on the other hand, you have been asked to submit a proposal, you can assume that the person requesting it knows something about the problem. You still need to describe the problem, however, as you see it. Do this to make it clear that you understand what you are being asked to do.

B3e: YOUR SOLUTION TO THE PROBLEM

State in a general way the solution you are proposing, what its benefits will be, and your qualifications for the project. Such things as previous experience on similar projects will help.

B3f: WHAT IS THE BODY OF THE PROPOSAL?

If your introductory material has described the problem, given some immediate background information, and proposed a general solution, then you need to convincingly develop all of the following relevant topics:

1. Need for solution.
2. The scope of your solution.
3. Cost and method of payment.
4. The feasibility of your solution.
5. Methods to be used.
6. Task breakdown--who will do what.
7. Time and work schedule.
8. Facilities available.
9. Personnel and their qualifications.
10. Urge to action.

B3g: WHO READS PROPOSALS?

Proposals should be aimed at persons who can make decisions--supervisors, managers, superintendents. Frequently, the cost of the proposed project will determine how far up the line your proposal will go. Your immediate supervisor will decide on some projects; others may go all the way to the president and the board of directors.

Unsolicited proposals made to outside companies or government agencies should be addressed to top managers. Solicited proposals should naturally be addressed to persons requesting them. In either case and to the extent possible, write so that interested non-technical persons can understand your material.

B3h: REMEMBER

An accepted proposal is in itself an engineering success. Without it, you will never have the opportunity to proceed.

ECE 411 Systems Analysis Project Proposal
Title: INTERFERENCE NOISE CONTROL IN AM RADIO SYSTEMS
Proposed by: Thomas N. Smith
Date: October 14, 1984

I.   Introduction and Background.   The quality of amplitude modulation (AM) radio reception in the presence of interfering noise leaves much to be desired, particularly when compared to frequency modulation (FM) reception.   A need exists to determine an easily implemented method of impulse noise rejection and overall improvement of the signal-to-noise (SNR) ratio in a typical AM portable radio.   There are many types of noise limiting circuits available which have been tested and are effective; however, most are complex and expensive.   This proposal involves the investigation of two very simple and inexpensive means of noise reduction: demodulation at the detector diode, and noise detection and elimination by blanking systems.   It is anticipated that at least 10 dB of SNR improvement can be obtained at some normal operating condition.   A seven week project is proposed to investigate these means of noise control in AM radio systems.

II.   Approach.   Through implementation and testing, circuit modification of the detector stage of a typical AM radio will be used to determine the most effective diode type and response time constant for noise rejection.   Simulated Gaussian noise from a General Radio laboratory noise source will be used to inject noise into an AM receiver.   Signal-to-noise ratio measurements will be made to identify any improvement in the noise reduction associated with the modified receiver.   To investigate performance in the presence of impulsive noise, laboratory generated impulse noise will be used to evaluate a simple detection and blanking circuit on the basis of measured signal-to-noise ratio improvement.   Various thresholds and blanking intervals will be investigated.   Based upon the laboratory measurements, an optimum receiver modification will be defined. Finally, using SPICE, a digital computer analysis of the optimum receiver circuit design will be made to analytically verify the observed SNR improvement.

III.   Schedule.   The following outline of critical events in managing this project is proposed.

        Week 1 - search literature.
        Week 2,3 - design circuits, order parts.
        Week 4 - build circuits.
        Week 5,6 - test circuit.
        Week 7 - prepare formal report.

IV.   Reporting.   An informal progress report will be submitted at the end of week 4 of the proposed project.   A final formal report will be submitted at the end of week 7.

CHAPTER B4
TECHNICAL REPORTS

B4a: <u>INTRODUCTION</u>

As an engineer you will often be asked to write progress reports, lab reports, and solutions to individual problems. The quality of your work will be determined in part from the appearance and content of the technical reports you write. In most instances, you will not be present when your report is read and evaluated, and therefore you will not have the opportunity to explain or clarify your work. Your reports must stand on their own. As with every communication situation, the analysis procedures presented in Section A should be reviewed and applied. The importance of determining your objectives and analyzing your reader(s) cannot be over emphasized.

Once you have completed an analysis of the intended readers, determine if the report must be on a pre-designed, or possibly serial-numbered, form or if your own format and style are satisfactory. Then utilize the following guidelines:

B4b: <u>GUIDELINES FOR WRITING REPORTS</u>

1. With any kind of report - short or long, formal or informal, start by analyzing the situation. Ask yourself <u>who</u> will be reading the report and <u>what</u> those readers need to know.

2. Decide what organization will best present your material. Consider your options. Do not, for example, force things into <u>chronological</u> order when <u>most</u> <u>important</u> <u>first</u> would be better.

3. Construct at least a rough outline before you start writing. Change it as necessary, but remember that organizing <u>after</u> writing is wasteful of time and effort. Think of your readers and your purpose as you design your report.

4. In an introductory statement orient your readers thoroughly. Tell them the purpose and the basic nature of the whole. Lead them into your report. The details come later.

5. Use headings and subheadings liberally. Try using statements rather than topics. Your reader should be able to see your organization in these statements.

6. Don't bury important ideas under details. Put much of your data in an appendix.

7. Concentrate more upon results than method unless for some reason your method is more important. In most investigative reports, readers are looking for what you found, not how you did it.

8. Draw conclusions and make recommendations if you have the authority and if your data warrant them.

9. Write an independent summary for every formal report you do. Inform rather than describe in the summary.

10. Provide a table of contents for reports longer than three or four pages. In effect, this will be your outline with page references.

11. Use an appendix for material not an integral part of the main message of your report.

12. When you have finished your rough draft, let it cool for a while--a day at least. Then revise and polish it.

13. Use tables, graphs, drawings, photographs, and other visuals to support and enforce your ideas.

14. Write clearly and as simply as you can, using plain, concrete words. Avoid indirect expressions and involved constructions.

15. Write grammatically, punctuate correctly to help your readers, and spell correctly.

B4c: PROGRESS REPORT

When someone gives you a problem to solve or a project to complete, he wants to know from time to time how your work is progressing. Are you on schedule? Are you within your budget? Have you run into unexpected problems? What do you plan to get done in the next period--month, quarter, year? The Progress Report is used to answer these questions.

Depending upon its importance and its readers, a progress report can be formal or informal. More often than not it is informal. It can be in the form of a letter, especially if it is going out of the company, or a memorandum if it is internal. At any rate, a series of progress reports should all have the same format.

The introduction to the first in a series of progress reports should identify the project and specify any particular methods and necessary materials. It should also give a completion date for the project.

The body describes in detail the present status of the project. The conclusion should make recommendations about changes in schedule, materials, and methods. It should state conclusions, if appropriate, and estimate future progress.

B4d: SAMPLE PROGRESS REPORT

TECHNICAL LETTER REPORT NO. 84

1 May 1985 to 31 May 1985

Contract No. 237421

DRD Line Item No. SE-5

Prepared for

CBA Engineering Area
Government Research Laboratory
4800 Fairview Drive
Pasadena, CA 91103

Prepared by

Department of Electrical and Computer Engineering
Clemson University
Clemson, SC 29631

Date of Issue: 10 June 1985

Approved by:

J.S. Waldrup
Principal Investigator
Clemson University
(803) 656-3375

Investigation of Accelerated Stress Factors
and Failure/Degradation
Mechanisms in Terrestrial Solar Cells
1 May 1985 to 31 May 1985

PROGRESS

1. Cell Testing

Unencapsulated Cell Testing -- The status of cells under test is indicated by the following table (with new measurements boxed):

Test	End Time	Cell Type	
		VA	RA
75 C B-T	4800 hours	4800 hours	2400 hours
135 C B-T	4800 hours	4800 hours	2400 hours
150 C B-T	2400 hours	2400 hours	2400 hours
Pressure Cooker	500 hours	500 hours	500 hours
85/85	2000 hours	2000 hours	2000 hours
Thermal Shock	40 cycles	40 cycles	----------
Thermal Cycle	40 cycles	40 cycles	40 cycles

Real Time Test Array (Encapsulated cells) -- Cells continue on test. The cells had a total of 16,428 hours as of 1 June 1985. The next scheduled downtime is at 17,520 hours on 15 July 1985.

Real Time Test Array (Unencapsulated cells) -- Cells continue on test. The cells had a total exposure of 5,408 hours as of 1 June 1985. The next scheduled downtime is at 9,760 hours on 1 October 1985.

Real Time Test Array (Thin film module) -- The series connected a-Si module that was installed in the outdoor test array had accumulated 1878 hours as of 1 June 1985. It was removed from the roof and tested in the laboratory under ELH illumination on 13 May 1985 after accumulating 1411.5 hours of outdoor exposure with the following results:

Voc	14.58 volts
ISC	39.76 ma
Pm	241.55 mw
Im	25.9 ma
Vm	9.33 volts
T	28.0 C

It is clear from a comparison of this data with that taken earlier that some degradation of the short circuit current has occurred. Physical changes can also be observed which are apparently due to a loss of back surface metallization despite the fact that the array is encapsulated.

Thin Film Cell Testing -- A 16-cell module, representative of a recently acquired lot of samples, is currently undergoing exploratory step stress testing. The module has completed three downtimes with the following results:

Stress Temp. (Degrees C)		Percent Decrease			
		Voc	Isc	Pm	FF
	entire module	-1.95	-0.78	-3.77	0.11
100	"good" cells	-0.09	-0.71	1.14	3.51
	"poor" cells	-20.27	1.26	-38.16	-60.67
	entire module	-2.49	2.42	-13.62	-1.88
120	"good" cells	-0.16	2.59	4.36	4.04
	"poor" cells	-18.65	1.26	-143.81	-107.88

Note that the poor cells show relatively large amounts of improvement as has been observed previously.

Additional samples of a-Si:H cells have been received and a preliminary test program has been outline for unbiased, non-illuminated, high temperature stressing.

2. Measurement Techniques

Thin Film Cells -- Spectral characteristics of a-Si:H cells made by two different manufacturers were measured and compared to the filtered photodiodes used in the simulated reference cell. Results are given in volts, but are unnormalized.:

Wavelength (nm)	photodiodes	Mfg. #1	Mfg. #2
700	6.75	5.83	3.42
600	2.69	7.79	7.91
500	1.38	6.12	4.55

The photodiode values are unamplified (or constant amplified) so that gain factors can be calculated from the ratio of the values. Thus,

	photodiode gain factors	
Wavelength (nm)	Mfg. #1	Mfg. #2
700	0.864	0.507
600	2.90	2.94
500	4.43	3.30

it is evident that the spectral response of the two types of cells is quite different.

A new circuit board has been constructed for use with the simulated array under DC simulator conditions. The board outputs all 5 of the photodiodes to the computer. Each of the photodiode outputs will be multiplied by an appropriate weighting factor and added in the computer to produce a value corresponding to one-sun.

The a-Si computer test program was modified to take second and fourth quadrant data and to display the reduced parameters, but has not yet been completely debugged.

No Auger analysis work was performed during the month, but some time was used for training. All equipment items have not yet been installed.

3. <u>Other</u>

A new graduate student, Judy Smith, has joined the program to replace Bill Jones, who completed his MS work in May and has accepted a job with Nippon Mfg. Co. He will be working in Japan for the next two years. One of Judy's main responsibilities will be failure analysis using the electron microscopes.

PROBLEMS

None

PLANS

-- Continue cell test program

-- Refine Auger analysis techniques for a-Si cell

-- Complete step stress testing and establish a-Si test schedule

-- Debug a-Si tester program

1. The purpose of a laboratory report is determined by the problem addressed: to find an unknown substance in a solution; to test the strength of a piece of steel; or to test the durability of a piece of plastic.

2. Each laboratory establishes its own organization for reports, but all reports contain the following:

    ### Introduction

    Here the problem, purpose, scope, equipment, and procedure are discussed to lead the reader into the report and prepare him for the more detailed material to follow.

    ### Body

    Here the data are presented, interpreted, and discussed. Remember that a tabulation of data isn't enough in most cases. Interpretation is necessary.

    ### Conclusion

    Here you draw your conclusion from the previous parts of the report and make recommendations.

3. Because of the critical nature of equipment and procedures, these two items are often emphasized more strongly than they might be in other types of reports.

4. If your report needs tables and graphs, be sure they are clearly titled, numbered, uncluttered, and easily read. They may be integrated into the text or placed at the end.

5. Use of passive voice is normal in laboratory reports, where the doer is not as important to the reader as what was done.

B4f: SAMPLE LAB REPORT

TITLE: Second Order Circuit Frequency Response

SUBMITTED BY: Thomas N. Smith

DATE: September 4, 1985

COURSE: ECE 411

PURPOSE: To investigate the magnitude and phase characteristics of the sinusoidal response of a second order electrical network.

PROCEDURE: The second order RLC circuit shown in Figure 1 was constructed in the laboratory. Using a sinusoidal signal generator to supply an input signal, an oscilloscope was used to measure the voltage amplitude gain over a frequency range from 1 Hz to 20 kHz. Using the network transfer function for this circuit, the input/output phase shift as a function of frequency was computed over this same frequency range. Using these data, a Bode plot description of the transfer characteristics for this network is presented.

PRESENTATION OF DATA: Data were collected on August 30, 1984, using an HP model 200CD sinusoidal signal generator as input and a Tektronix 2215 dual trace oscilloscope to measure the peak-to-peak voltage at the input and the output of the circuit investigated. The circuit was implemented using R, L, and C substitution boxes from the equipment cabinet in Riggs 200. Recorded data are presented in Table 1. The input voltage was maintained at 1 volt so that the voltage GAIN computed as the ratio of input to output is simply the output voltage value. Table 1 includes the network gain computed in decibels (dB) as $20\log_{10}$GAIN and the network phase shift in radians computed as the ARCTAN of the ratio of the imaginary part to the real part of the network transfer function, given by

$$H(jw) = \frac{wRC}{wRC + j[1-(w/w_o)^2]}$$

where $w_o = 1/\sqrt{LC}$ is the resonant frequency of this circuit.

DISCUSSION OF RESULTS: Figure 2 presents the Bode plot, including both the magnitude and phase response, for the second order network under investigation. This circuit can be categorized generally as a bandpass filter with a gain of unity at the resonant frequency, w_o=10000 radians/sec (1592 Hz). From the phase plot it can be observed that the phase shift of the output relative to the input is near 90° for frequencies below the resonant frequency, a rapid phase shift change through zero occurs near the resonant frequency, and a -90° phase shift is present at frequencies above the resonant frequency. As can be seen from the magnitude gain plot the 3 dB bandwidth associated with this network is approximately 1000 radians per second (159 Hz). Within this 3 dB bandwidth the network can be categorized as a linear phase network. The Q of this circuit, defined as the resonant frequency divided by the 3 dB bandwidth, is 10.

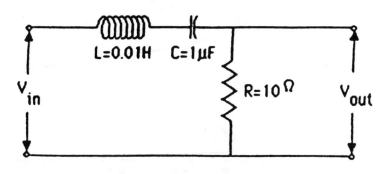

Figure 1. RLC Network

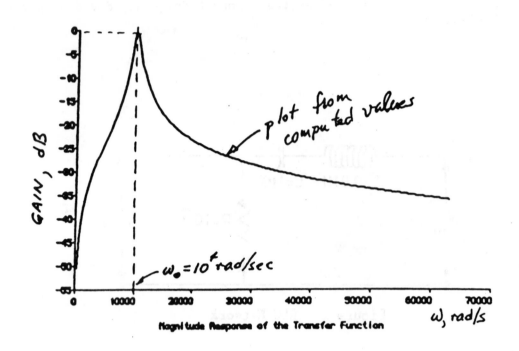

Magnitude Response of the Transfer Function

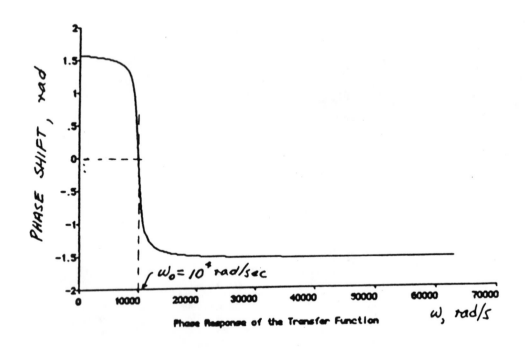

Phase Response of the Transfer Function

Figure 2. RLC Network Bode Plot

TABLE 1.

RLC Circuit Response Data

Computed columns: Vout, GAIN(dB), φ (rad), Uin Vout GAIN(dB)

f (Hz)	rad/s	Uin (volts)	Vout (volts)	GAIN(dB)
100	628.32	1	0.006	-44.4
500	3141.6	1	0.035	-29.1
900	5654.9	1	0.08	-21.9
1300	8168.1	1	0.24	-12.4
1700	10681	1	0.6	-4.4
2100	13195	1	0.18	-14.9
2200	13709	1	0.1	-20.0
2600	16221	1	0.08	-21.9
3300	20735	1	0.06	-24.4
3700	23248	1	0.05	-26
4100	25761	1	0.05	-26
4500	28274	1	0.04	-27.9
4900	30788	1	0.04	-27.9
5300	33301	1	0.03	-30.4
5700	35814	1	0.03	-30.4
6100	38327	1	0.03	-30.4
6900	40841	1	0.025	-32.0
6900	43354	1	0.025	-32.0
7300	45867	1	0.02	-33.9
7700	48381	1	0.02	-33.9
8100	50894	1	0.02	-33.9
8500	53407	1	0.02	-33.9
8900	55920	1	0.017	-35.0
9700	60947	1	0.015	-36.5
10000	63460	1	0.015	-36.5
10300	65973	1	0.015	-36.5
10700	68487	1	0.015	-36.5
11300	71000	1	0.014	-37.1
11700	73513	1	0.0145	-38.1
12100	76027	1	0.0125	-38.1
12500	78540	1	0.011	-39.1
13300	83053	1	0.011	-39.1
13700	86080	1	0.011	-37.2
14100	88593	1	0.01	-40
14100	91106	1	0.01	-40
14600	93619	1	0.01	-40
15300	96133	1	0.01	-40
15700	98646	1	0.01	-40
16100	101159	1	0.01	-40
16000	103672	1	0.01	-40
16900	106186	1	0.01	-40
17700	110699	1	0.009	-40.9
18100	113726	1	0.009	-40.9
18600	116239	1	0.008	-41.9
18900	118752	1	0.008	-41.9
19300	121265	1	0.008	-41.9
19700	123779	1	0.008	-41.9
20100	126292	1	0.008	-41.9

(Computed values table)

f (Hz)	rad/s	Uin Vout (volts)	Gain (dB)
1400	8796.5	.32	-9.9
1410	8859.3	.38	-8.4
1420	8922.1	.4	-7.9
1430	8985	.4	-7.7
1440	9047.8	.45	-6.9
1450	9110.6	.475	-6.5
1460	9173.5	.5	-6.0
1470	9236.3	.525	-5.6
1480	9299.1	.55	-5.2
1490	9361.9	.6	-4.4
1500	9424.8	.65	-3.7
1510	9487.6	.675	-3.4
1520	9550.4	.725	-2.8
1530	9613.3	.775	-2.2
1540	9676.1	.825	-1.7
1550	9738.9	.875	-1.2
1560	9801.8	.925	-0.7
1570	9864.6	.95	-0.4
1580	9927.4	.99	-0.1
1590	9990.3	1.0	0
1600	10116	1.0	0
1610	10179	.975	-0.2
1620	10179	.95	-0.4
1640	10242	.9	-0.9
1650	10367	.85	-1.4
1660	10430	.8	-1.9
1670	10493	.75	-2.5
1680	10556	.725	-2.8
1690	10619	.675	-3.4
1700	10681	.65	-3.7
		.60	-4.4

* COMPUTED VALUES

B4g: GUIDELINES FOR PRESENTING SOLUTIONS TO PROBLEMS

As an engineer you should realize that there are two equally important tasks in completing a problem assignment. First, you must determine a solution for the problem, and second, you should record the solution in a neat, concise, and logical format that is easy to read and comprehend. This should be a distinguishing mark of an engineer's work. Some guidelines are as follows:

1. Most solutions are presented on 8-1/2 by 11-inch paper commonly known as engineering problems (or calculation) paper.

2. Do not crowd the presentation.

3. Use lettering and do not write with script.

4. Use a well sharpened 2H lead so that the marks are dark enough to be easily read.

5. Be liberal in the use of sketches.

6. Begin each problem solution at the top of a blank page.

7. Separate the presentation into at least four areas in the following order:

   a. Course or project identifier, problem identifier, page number, date, and your name.

   b. Description of problem, including sketch and data.

   c. Statement of what is required.

   d. Solution.

8. Clearly indicate units.

9. Mark answers distinctly.

10. Number pages.

| ENGR 180 | PROBLEM 5.6 | SMITH, L.B. | $\frac{1}{1}$ |

GIVEN:

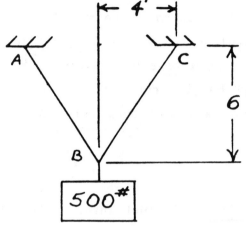

A 500 LBF WEIGHT IS SUSPENDED BY A
FLEXIBLE CABLE A-B-C.

REQUIRED: FORCE IN CABLE DUE TO
THE 500 LBF WEIGHT.

SOLUTION:

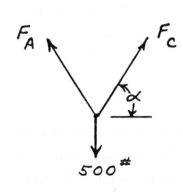

$$F_A = F_C$$

$$TAN \; \alpha = \frac{6}{4}$$

$$\alpha = 56.3°$$

$$\Sigma F_y = 0$$

$$2 \; F_C \; SIN \; \alpha = 500 \; LBF$$

$$F_C = \frac{500 \; LBF}{2 \; SIN \; \alpha}$$

$$F_C = \underline{300 \; LBF} \qquad \text{ANS}$$

CHAPTER B5
TECHNICAL ARTICLES

B5a: INTRODUCTION

As an engineer you will have ideas and projects that you wish to share with members of your profession.   One way of doing this is by writing articles for technical journals such as Civil Engineering.   As a student, you will be encouraged to submit articles for publication; as a practicing engineer you may be required to write them.   If these are done well, they reflect favorably upon both the writer and the company.

B5b: PREPARATION

Normally the article will be of wider interest than the technical report.   It is, therefore, written for a more general audience.   The usual steps are gathering information, planning your paper, writing, and revising.

B5c: Gathering Information

In the gathering and creative selection of information, there are three major sources: personal and professional experience, library sources, and original research.   In this phase, first consider the message you wish to present, then your audience (readers and their background), next the purpose and scope of your article.   For more specific details, see paragraph A1a for objectives and A1b for audience analysis.

B5d: Writing Phase

The first step in this phase is planning.   Your plan should be based on a mental path that leads the reader from the introduction through the body and the conclusion.   Take whatever relaxed, exploratory time necessary to develop the information and concepts you wish to present; generously use headings and subheadings to guide and keep the readers oriented as they progress.   The remainder of this chapter reviews the major parts of a technical article.

B5e: Introduction

It should present four things: the subject with any necessary background, the purpose, the scope, and the plan of development.

B5f: Body

The body develops the information to be presented. To do this you may describe, narrate, and even argue a point. Your main purpose, however, is usually expository - to explain. Visuals - photos, drawings, charts, graphs - are often helpful in supplementing your writing.

B5g: Conclusion

Draw whatever conclusions are warranted by the body of the paper. If no definite conclusions are in order, then a simple summary may be all you need. Above all, do not introduce new material here.

B5h: Abstract/Summary

The terms _abstract_ and _summary_ are often used interchangeably. However, two types of abstracts can be distinguished--descriptive and informative. The descriptive abstract reflects _what the original is about_; the informative abstract tells _what the original contains_. The first of these usually takes no more than a brief paragraph. Sometimes it is done in a sentence or two. The informative abstract gives the major facts on which conclusions are based, the conclusions, and any recommendations made. This type of abstract can be called a summary. It is appropriate for technical articles and reports. Normally written last, it appears at the beginning of the article or report.

B5i: GENERAL GUIDELINES

A basic rule of technical communications is to give the audience the most information of value while requiring the minimum time and effort. The following guidelines will help you do this. Keep them in mind especially when you are revising.

1. Use topics and subheadings liberally. They will help your readers follow your ideas and retrieve information.

2. Be sure the article contains everything the reader needs.

3. Ask yourself how much you can remove without reducing the readers' understanding or their ease of reading. For example, have you put in things to impress - big words, or five words when two would be better.

4. Strive to be so clear that you cannot be misunderstood.

5. Be sure your grammar and punctuation are correct.

6. Use transitional words and phrases to help your readers.

B5j: Documenting Your Sources

Naturally, you will use the works of others in your research. Be sure to give credit to these sources by proper footnotes and bibliography. Since methods vary greatly from one periodical to another, follow the method used in the one to which you submit your article. If, however, you duplicate more than 250 words (even scattered quotes), or more than 5 percent of an original work, you should obtain written permission from the publisher.

B5k: Other Format Requirments

Engineering societies such as IEEE, ASME, and others provide author's guides and often an author's kit. These include information which will help you prepare your article in the proper format and thus avoid unnecessary re-write and delay.

B5l: REMEMBER

As you read the article below, look for the following: an introduction which leads you into the article, a body which develops the main ideas in a logical sequence, and a conclusion which summarizes the major points.

B5k: <u>SAMPLE TECHNICAL ARTICLE</u>
THE ECONOMIC ROLES OF THE ENGINEER

Abstract

This article reviews the numerous engineering functions required in defining, designing, developing, manufacturing, and marketing of a new product.

This is a <u>descriptive</u> abstract.

Introduction

Engineers in both private industry and government are instrumental in meeting the needs of mankind. This article describes a product development by private industry but would also apply to a government project if references to a "sales" versus cost were replaced with "public benefit" versus cost. In a private company the engineer's role normally starts after managers have determined the need for a product and the feasibility of manufacturing it. In this brief article we shall look at the engineer's role in research and development, designing a product, developing and testing a prototype, establishing a pilot production line, and eventually a full-production line. When the product is highly technical, engineers may also work with sales and marketing personnel in defining the need and applying technical support to the finished product.

This introduction forecasts the structure of the article.

Determining the Need

Consider the block diagram in Figure B5-1. It presents a simplified view of the workings of a private company. The economic process begins with a need. Someone or some group needs something. The need can range from the fundamental, such as a source of pure water for drinking, to the somewhat frivolous, such as a dog food with a higher nutritional content. The company itself may have induced the need, perhaps through a clever advertising campaign. A company may even have made a product and then induced a need for it.

Subheadings help the reader follow the ideas. If done carefully, these subheadings form an outline.

Feasibility

In any event, once the need (or potential need) is apparent, the company's management must estimate the potential market size, the cost of manufacturing, and the eventual selling price. Recognizing that profits equal total sales revenue minus total cost, they must

The chief organizing principle in this article is chronolo-

65

answer the question: is it economically feasible? If the answer is no, the process stops, as indicated in Figure B5-1, because the company has concluded that it cannot satisfy the need profitably. Profit is essential if the company is to grow and provide jobs to support the economy. If the answer to the question is yes, the company begins a process that may ultimately lead to a new product. (We say "may" because many projects designed to develop new products fail for a variety of reasons.) And at this time a large company's research and development department (usually shortened to R and D) enters the picture.

gical. It progresses in a series of logical steps.

Using an occasional <u>and</u> to start a sentence is all right, but it should not be used this way very often.

The Research Team

Once the company decides to proceed, they assign a research team (in some cases the team has only one member) the task of determining whether scientific principles needed to develop the proposed product are known. The research team searches the scientific and engineering literature as well as the patent registry. If the scientific princples are either unknown or inadequately understood, the company must conduct the necessary research to sufficiently understand these principles.

Note use of time order in giving the steps.

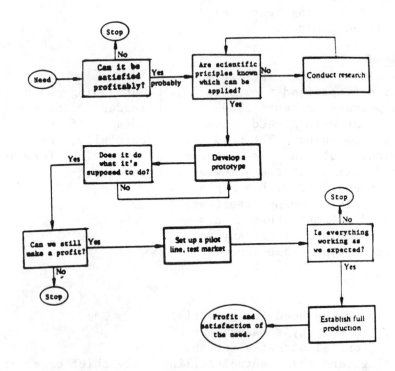

Figure B5-1. Decision making in the economy – a simplified view.

It is during this research phase that an engineer generally becomes involved in the overall development and economic justification process. Once the engineer finds a principle that has no serious drawbacks or flaws, this process proceeds to the next phase. If appropriate, the company normally seeks a patent. We should point out that research costs money, and [sometimes] when a company does significant work without making progress [toward a solution], the company's mangement decides not to spend any more [money] on the project. Figure B5-1 does not explicitly show this possibility, but it is an important consideration at each step in the process.

[ ] Brackets indicate unnecessary words. You may find this in other places.

Development of a Prototype

The next phase is development. A prototype of the final product is assembled and tested to see if it meets all the necessary conditions. These include performing as required, meeting cost targets, and satisfying all applicable regulations and laws. If a given prototype is inadequate, another must be developed and tested. Again, if this process is repeated too many times without progress, the project may be terminated. Once a successful prototype has been developed, the company's marketing personnel again investigate and assess the profit potential [of the product]. This reassessment is essential if the research and development phases have consumed significant amounts of time.

Note use of transitional phrase "the next phase."

Use small words when available: stopped or ended for terminated.

Pilot Production Line

Next, the company normally establishes a pilot production line, produces limited amounts of the product, and test markets it. Most of the design details must be determined and documented at this time. If the market tests are positive, the company commits money to the construction or purchase of tools, equipment, and space necessary for full-scale production and national or even international sales. Periodically, they review the performance of each product line to ensure that it remains profitable.

The Engineer's Role

Engineers fit into this economic flow at a number of points. Research engineers search for new scientific principles and a more complete understanding of them. Development engineers exploit these principles to produce prototypes. Product design engineers perform the basic task of determining and documenting product details. Industrial and manufacturing engineers set up the pilot production lines and specify the necessary equipment and space. Plant engineers see that the design and operation of a new production facility are successful. In addition, as stated earlier, highly technical products often require R & D engineers to help define the need and application engineers to assist the user of the product.

Conclusion

From this brief examination of the economic roles of engineers, it can be concluded that engineers normally enter the process after managers have decided upon the need for a product and the feasibility of producing it profitably. Engineers initially determine workable principles for making the product, and then design, produce, and test a prototype. Other engineers document the design and set up a pilot production line. Finally, if full production proceeds, engineers design and oversee the new production facility. The role of the engineer is thus an essential key in the progress of our society.

Article is summarized. No new material introduced.

CHAPTER B6
RESUMES

B6a: ANALYZE FOR BETTER COMMUNICATIONS

When you apply for a job you will need to write a resume and a letter of application. These two normally are sent together to prospective employers. See Chapter B2 for a sample letter of application and review the analysis procedure in Section A.

Your analysis should make these assumptions about the people who will read your letter and look at your resume:

1. They are intelligent people who have seen many resumes and letters, some excellent, some fair, some unacceptable. Thus they will be able to compare yours with others.

2. Their company is successful because they hire the best people they can find. This means people who can help the company.

3. They are busy. They have no time to waste on badly written lengthy letters or on poorly organized resumes.

4. They will have many applications to look over. You have competition.

B6b: THE RESUME

The resume is a condensed autobiography in outline form stressing your qualifications. Name, address, and telephone number appear at the top. Personal data, education, experience, and a list of references make up the usual resume.

B6c: HOW TO ARRANGE

Consider arranging these topics in the order that will best emphasize your strong points. For example, you might put your experience first if you have had several jobs relevant to the position for which you are applying. Putting it first will call attention to it and separate you from those who have had little or no experience.

On the other hand, if education is your strongest qualification, as it will be for many, then put it first. Do not waste your reader's time by listing numerous courses that everyone in a given curriculum would take. Stress things that make you a little different--language skills, aptitude for writing or speaking, courses with special projects--computer courses, or whatever makes you special.

Personal data is not as important as it appears to be when put first. Almost all applicants with college degrees will be close to twenty-two years old, of average height and weight, and in good health. The word sick is hardly ever found in the blank opposite health.

B6d: REFERENCES

You can find examples that say "references furnished upon request." Consider, however, the advantages of listing at least three references. First, the prospective employer does not have to call or write you to request them. Secondly, having the list on your resume suggests, if subtly, that it would be easy to call one or more of them. Finally, there is a good possibility that prospective employers will know some of the persons listed, especially if you list one of your professors.

B6e: SHAPING YOUR RESUME

Look at a number of different types of resumes before deciding on the form you will use. Notice that some will state the applicant's professional goals. Some will stress college activities--especially those showing leadership roles. If you include these, limit them to those most relevant to the job you are after. For example, if the job you want is in research, it would do little good to point out activities better suited to sales.

B6f: APPEARANCE

Think of your readers. Leave some white spaces. Use plenty of headings and subheadings. Complete sentences are not necessary, and you may use accepted abbreviations. If you can get it all on one page, do so, but do not crowd it. Use two pages and space the items for easy reading.

<u>B6g</u>: <u>REMEMBER</u>

The resume should not be sent without a letter of application. In fact, the resume should be attached so the letter is read <u>first</u>. See Chapter B2 for key features and an example of this very important letter. The letter, and a well prepared resume, is your first and sometimes only key to the "interview door."

B6h: SAMPLE RESUME

Michael J. Brunig

College Address

314 Newman Hall
Clemson University
Clemson, SC 29631
Tel: 803-656-8031

Home Address

821 Cedar Cr.
Greenwood, SC 26803
Tel: 803-921-7653

Experience

Summer 1984 - Worked in robotics department for Mr. Larry Stone.
              Used McCrory robots to paint machinery.
Summer 1985 - Worked at Kusoi Robotics in Sendai, Japan. Did some
              drafting and some designing. Worked for Mr. Tasaki
              Yakusoto.

Education

B.S. Degree in Mechanical Engineering, Clemson University - Date of
     graduation May 15, 1985.

     Special courses: Robotics Design - (6 hrs.)
                       Computer Engineering - (3 hrs.)
                       Japanese - (6 hrs.)
                       Technical Writing - (3 hrs.)

Honors and Activities

Milliken scholar - two years
R.F. Poole scholar - four years
Dean's List - 5 semesters (GPR 3.75 through 7 semesters)
Mechanical Engineering Society - (President junior year)
Class Vice-President (sophomore year)

Personal

Age: 21
Height: 5 ft., 11 in.
Weight: 170 lbs.

Marital status: single
Health: good
Birth date: Jan 9, 1965

References

Dr. Tom Abernathy
Dept. of Mech. Engr.
Clemson University
Clemson, SC 29631
Tel: 803-656-9031

Ms. Jane Cason
Cason & Cason, Attorneys
Abbeville, SC 29691
Tel: 803-421-7036

Mr. Larry Stone
Platt-Saco Lowell
Easley, SC 29169
Tel: 803-651-9824

CHAPTER B7

GRAMMAR AND PUNCTUATION: A SHORT GUIDE

B7a: ABBREVIATIONS

Abbreviations are widely used in technical writing. Be sure, however, that they will be understood by your readers. When in doubt spell the complete word. Avoid using etc. in formal writing since readers may not know what you are including.

B7b: AGREEMENT

Make your verb agree with its subject. This problem occurs most often when a subject and verb have intervening words.

Wrong: A list of absent members are attached.

Right: A list of absent members is attached.

Make your pronouns agree with their antecedents. In the above sentence their correctly agrees with the word pronouns.

Wrong: Everyone in the class raised their hand.

Right: Everyone in the class raised his hand.

Use the nominative case for subjects, the objective for objects.

Wrong: Dr. Thompson enjoys teaching the computer to whomever wants to learn.

Right: Dr. Thompson enjoys teaching the computer to whoever wants to learn.

In this sentence whoever is the subject of the verb, not the object of the preposition to.

Wrong: The task was divided between he and Greg.

Right: The task was divided between him and Greg.

Both him and Greg are objects of the preposition between.

B7c: AMBIGUITY

This word refers to double meaning or vagueness.

Wrong: John told James that Bill didn't like him.

Right: John said to Jim, "Bill doesn't like me."

B7d: CAPITALIZATION

Capitalize proper names, the first word of every sentence, the first word of each item in an outline, the first and every important word in titles of books, magazines, newspapers.

Do not capitalize points of the compass, the seasons, names of academic studies unless they are specific courses: Philosophy 306, psychology, engineering, mathematics.

B7e: CHOPPY SENTENCES

Too many short sentences can cause problems because they are monotonous to read and because the writer makes the reader determine how the choppy sentences are related. The tendency to write a choppy style is sometimes referred to as the Dick and Jane syndrome or the Jack Webb (Dragnet) syndrome. Use an occasional short sentence for emphasis.

B7f: COHERENCE

Make your ideas stick together by logical development and by using transitional words carefully. These words can come within sentences as well as between them. However, moreover, consequently, as a result, nevertheless, first, second are examples.

B7g: DANGLING MODIFIERS

Opening verbal or prepositional phrases will appear to modify the first noun or pronoun following them. Be sure it is the right word.

Wrong: Having rear end trouble, I was driving slowly down the highway.

Right: Because my 1978 Plymouth was having rear end trouble, I was driving slowly down the highway.

B7h: FRAGMENTS

Do not punctuate subordinate clauses or phrases as if they were sentences.

Wrong: While the machine was out of order.

Right: While the machine was out of order, production was down
by six percent.

B7i: PARAGRAPHS

Every paragraph should have a topic sentence, usually the first one. The other sentences should stick to the topic and develop the main idea by details, examples, logical argument, comparison and contrast. Paragraph unity means that all sentences help develop one idea.

B7j: PASSIVE VOICE

Use the passive voice when the doer of the action is unknown, unimportant, or not to be mentioned for some reason. Laboratory reports are correctly written in the passive voice.

Active: I heated the liquid to 100 degrees centigrade.

Passive: The liquid was heated to 100 degrees centigrade.

Use the active voice to emphasize the doer of the action.

Passive: It is recommended by the architect that steel reinforced
concrete be used.

Active: The architect recommended that steel reinforced concrete
be used.

B7k: PUNCTUATION

Punctuation properly used helps you say what you mean to say. Improperly used, it slows down communication and, at times, stops it. Learn to punctuate by rule, not by intuition. Do not overpunctuate. Overuse of the comma is a big offender.

B7l: The Comma

The comma is used in the following ways:

(1)   Between two independent clauses (complete sentences)
when they are joined by a coordinating conjuntion--and,
but, or, nor, for, yet:

Right: James ran the movie projector, but he did not
know how to splice film.

Wrong: James ran the movie projector, and spliced film.

(2) After long introductory phrases and clauses:

Right: Even though we had bought the television set used and had had it for three years, the company repaired it for their cost.

Wrong: When Mary left, he cried for two hours.

(3) With words, phrases, and clauses in series:

Right: The letter asked students to send name, address, sex, and housing requirements.

Wrong: The letter asked students to send name, address, sex and housing requirements.

(4) To set off nonrestrictive elements (clauses, phrases, appositives):

Right: Mr. James McCrimmon, who just took his place on the jury, teaches Latin at the academy.

Wrong: Mr. James McCrimmon who just took his place on the jury teaches Latin at the academy.

B7m: The Semicolon

The semicolon is used as follows:

(1) Between two independent clauses not joined by and, but, or, nor, for, yet:

Right: James graduated from high school; he attended college in Charleston.

Wrong: James graduated from high school, he attended college in Charleston.

(2) Between two independent clauses joined by a conjunctive adverb--however, moreover, consequently, nevertheless, and such words:

Right: Mistakes were made in the calculations; nevertheless, we were given the contract.

Wrong: Mistakes were made in the calculations, nevertheless, we were given the contract.

B7n: The Colon

The colon is used as follows:

It is used to introduce a long formal list. Be sure to use a
complete sentence before the colon.

>   Right: The following men will report to headquarters as
>   soon as possible:
>   Wrong: Our reasons for refusing the contract are:

B7o: The Apostrophe

The apostrophe is used to form possessives and to indicate
contractions.

>   Right: It's all right with me if you change your plans.
>   Wrong: The cat caught it's tail in the trap.
>   Right: The cat caught its tail in the trap.

The possessive of personal pronouns has no apostrophe: his,
hers, its.

CHAPTER B8
ERROR MESSAGES - WRITTEN COMMUNICATIONS

B8a: ANALYZE FOR BETTER COMMUNICATIONS

A critical analysis of your writing will provide the important feedback required for improvement.

B8b: ERROR MESSAGES

Example: "DC6c" means "choice of words inappropriate-too fancy."

Diction Related

DC Choice of words

DS Size of words

DU Use of words

Sentence Related

SC Conjunctions

SG Grammar

SL Length

SM Modifiers

SP Punctuation

SS Structure

SV Verbs

Paragraph Related

PC Coherence

PD Development

PL Length

PS Subject

PT Transition

Introduction Related

IB Background information

IC Content of introduction

IP Purpose of document

IS Scope of document

Organization Related

OD Division of topics

OI Ideas

OP Principles of organization

OS Sequence of organization

<u>Tone</u> <u>or</u> <u>Style</u> <u>Related</u>

TC Choice of style

TF Formality of style

TV Voice style

<u>Graphics</u> <u>Related</u> (Any figure or illustration)

GC Choice of graphic methods

GN Number of graphcs

GP Placement of graphics

GQ Quality of graphics

GR Readability of graphics

GS Size of graphics

B8c: <u>ADJECTIVE CODE</u>

0 Superfluous

1 Marginal

2 Distracting

3 Awkward

4 Uninteresting

5 Inadequate

6 Inappropriate

7 Improper

8 Excessive

9 Offensive

B8d: <u>CORRECTIVE ACTION CODES</u>

<u>Diction</u> <u>Related</u> - <u>D</u> Prefix

a. Words too technical

b. Words too elementary

c. Words too fancy

d. Words too abstract

e. Words too general

f. Words repeated too often

g. Words misspelled

h. Abbreviations not defined

<u>Sentence Related</u> - <u>S</u> <u>Prefix</u>

a. Fragments punctuated as sentences

b. Too many short, choppy sentences

c. Dangling or misplaced modifiers

d. Too much coordination or use of "and"

e. Overuse of passive voice

f. Long, loosely constructed sentences

g. Overuse of "I" and "we"

h. Overuse of "you"

i. Prepositions without objects

j. Tense

k. Overuse of commas

<u>Paragraph</u> <u>Related</u> - <u>P</u> <u>Prefix</u>

a. Topic not clearly expressed

b. Paragraph not unified

c. Paragraph undeveloped

d. Sentence flow illogical

e. Transitional words missing

f. Misuse of transitional words

g. Too many topics

h. Repetition

<u>Introduction</u> <u>Related</u> - <u>I</u> <u>Prefix</u>

a. Too technical

b. Too elementary

c. Poorly planned

d. Illogical

e. Incoherent

f. Too long

g. Too short

h. Unclear

<u>Organization</u> <u>Related</u> - <u>O</u> <u>Prefix</u>

a. No organization principle evident

b. Important ideas given subordinant positions

c. Order of importance should be used

d. Chronological order should be used

e. Major divisions unclear

f. Document too lengthy

g. Document too short

<u>Tone</u> <u>or</u> <u>Style</u> <u>Related</u> - <u>T</u> <u>Prefix</u>

a. Too formal for purpose and readers

b. Too informal for purpose and readers

c. Too indirect - too much passive voice

d. Too forceful or demanding for the circumstances

e. Not forceful enough for the circumstances

f. Too pompous, inflated, or wordy

<u>Graphics</u> <u>Related</u> - <u>G</u> <u>Prefix</u>

a. Figure too far from discussion

b. Figure too cluttered

c. Figure not clearly labeled

d. Figure needs further explanation

e. Figure too small

f. Figure too large

g. Figure not suitable style or quality

h. Figure needs caption

i. Lettering too small

j. Lettering too large

k. Lettering not suitable style or quality

CHAPTER B9

REFERENCES

FOR

WRITTEN COMMUNICATIONS

Brusaw, C.T., Gerald J. Alred, and Walter E. Oliu. Handbook of Technical Writing. New York: St. Martin's Press, 1982.

Gajda, Walter J. Jr., and William E. Biles. Engineering: Modeling and Computation. Boston: Houghton Mifflin Company, 1978. (Pages 12-14 are basis for B5k: Sample Technical Article. Permission granted.)

Hodges, John C., and Mary E. Whitten. Harbrace College Handbook. 9th ed. New York: Harcourt Brace Jovanovich, 1982.

Lannon, John M. Technical Writing. 3rd ed. Boston: Little, Brown and Company, 1985.

Tichy, H.J. Effective Writing for Engineers, Managers, Scientists. New York: John Wiley and Sons Inc., 1966.

Turner, Maxine T. Technical Writing: a Practical Approach. Reston, VA: Reston Publishing Co., 1984.

SECTION C
ORAL COMMUNICATIONS FOR THE ENGINEER

As an engineer, you will need the best possible oral communication skills. These skills are critical because most of today's <u>technical</u> activity must be reasonably understood and approved by <u>non-technical</u> people.

Whether in conversations, meetings, or formal presentations you must learn to gage your audience and be sure that you are speaking neither "up" nor "down". Refer to Section A and particularly subparagraph A1c on in-process analysis.

Section B covered the important facets of written communication, but you cannot escape the fact that many influential people throughout your career will be "listeners", not "readers."

CHAPTER C1

CONVERSATION

C1a: ANALYZE FOR BETTER COMMUNICATIONS

Conversation is the opening chapter of Oral Communications because conversation, rather than formal meetings or presentations, will be your most important information vehicle. The fundamentals described are naturally applicable to the more specific topics discussed in subsequent chapters. In your profession, you are likely to speak or hear more than 100 words for every word that you read or write. This important subject is divided into the following segments:

- Guidelines for listening.
- Guidelines for speaking.
- Conversational Check-list.
- Conversation with co-workers.
- Conversation with superiors.
- Conversation with subordinates.
- Conversation with clients.

C1b: GUIDELINES FOR LISTENING

You must concentrate on the technique of listening in order to master the technique of talking. As infants, we learn to talk by listening but tend to forget the process as we mature. Remember that communication is defined as what your listener hears, not what you say. Therefore, listening becomes a great deal more than just giving another person his turn to talk. Effective listening during a technical conversation will be one of your most important sources of knowledge, and a truly good listener can greatly enhance the initial ideas of the person speaking.

Here is a list of things to do and not to do for effective listening:

- Listen with the intent of understanding, not evaluating.
- Periodically restate what you think you have heard.
- Try to appear thoughtful and attentive. The speaker's next sentence may be the one that you really need to hear.

- Do <u>not</u> fake attention to the speaker; it's rarely successful.
- Do <u>not</u> listen for facts only; grasp the ideas of the speaker.
- Do <u>not</u> let emotional words or unmetered adjectives break your concentration.
- Do <u>not</u> waste time between the speaker's words. Speech is approximately 125 words per minute, but your understanding may be a hundred times faster. Use these gaps to analyze what is being said rather than day dreaming or composing your own next statement.

C1c: GUIDELINES FOR SPEAKING

You should recognize that <u>full</u> understanding is possible only between persons who are perfectly like-minded; therefore, your efforts can be only to achieve the highest possible percentage of understanding in each situation. Do not assume that anyone will have exactly the same understanding that you have, but improve your chances by having the following:

- Something to say.
- Fluent command of the language.
- Originality. Be able to place things in a view not commonly seen.
- Confidence. You must not fear failure.
- Ability to recognize patterns of disagreement.
- Ability to criticize constructively.

C1d: Something To Say

This chapter concerns technical conversation, and as an engineer you must "engage brain before opening mouth." If your lack of knowledge precludes comment, then pose an occasional question to help others develop their ideas. However, do not let your engineering training cause you to avoid "small talk" and become non-vocal in social situations. Make a clear distinction between a technical and social conversation and adjust accordingly.

C1e: Fluent Command of the Language

You should develop your vocabulary, both technical and non-technical. Be curious about words and remember that they have different and sometimes conflicting meanings; this is certainly true in the English language. As an extreme example, to "table" a proposal means "forget it" to an American but means "discuss it" to an Englishman. Imagine the potential disaster of General Montgomery asking General Eisenhower to table the invasion of Europe.

Make sure to use alternate words, tone of voice, posture, and gestures to clearly convey the meaning of what you are saying.

C1f: Originality

Your greatest contribution to a conversation is originality. If you have listened well, then you may have detected areas of uncertainty from some speakers and areas of misunderstanding between others. Originality allows you to rephrase ideas and facts from another point of view and thus enhance the understanding of an entire group.

C1g: Confidence

Confidence is a personality trait that goes beyond technical conversation, but you must make every effort to appear confident if you are to contribute. The timid are not often heard. However, do not confuse confidence with bluffing or pretending to have knowledge that you do not have. A simple statement such as "I do not know but will look into the matter" can be said with great confidence. Above all, do not let confidence become arrogance.

Accept criticism of your ideas and suggestions without assuming that it is criticism of your character or motives.

C1h: Patterns Of Disagreement

As a conversation proceeds, patterns of disagreement develop among the participants. Your contribution to the conversation should always be aimed at converting all patterns to a pattern of inquiry. The typical patterns are as follows:

- Inquiry: "Please explain what led you to that conclusion."
- Incredulity: "How did you ever get that strange idea."
- Amusement: "Surely you aren't serious?"
- Suspicion: "What's in this for you?"
- Dismissal: "I don't want to hear anymore about it."

C1i: Constructive Criticism

Always confine your disagreement to ideas or issues and carefully avoid attacking the character or motives of others. This is not always easy but must be attempted. A good rule to remember is that criticism is accepted in direct proportion to the humility of the critic. You must say something like "Our data is admittedly incomplete; do we need further tests?" and not "Your results are all wrong, Joe."

C1j: CONVERSATIONAL CHECK LIST

Take time to analyze your conversational behavior. You should be able to answer "no" to the following seven questions:

1. Do you usually dominate the conversation?
2. Does your voice rise in pitch and volume as the conversation progresses?
3. Do you often lack information possessed by others in the conversation?
4. Do you antagonize others or become argumentative?
5. Do you find clear expression difficult or forget names and facts?
6. Do you take too long to get to the point?
7. Do you feel ill at ease among strangers or your superiors?

Obviously, any "yes" answers deserve your attention and a conscious effort to correct.

C1k: CONVERSATION WITH CO-WORKERS

Most technical conversations throughout your engineering career
will be with co-workers. A co-worker is anyone outside of your
particular chain of command. This naturally includes peers, the people
on the same organizational level as yourself, but also includes people
at organizational levels both above and below you. Here are a few
rules to follow in conversations with co-workers:

C1l: Peers

Conversations with your peers need not be reported or
summarized to anyone else in the organization. However, you must
individually "calibrate" each of your peers as to what they will
do with your words. As an engineer, you sometimes have power that
exceeds your organizational authority, and one of your
non-engineering peers might use a conversation as a basis for an
inappropriate action.

C1m: Lower Level Co-worker

Conversations with co-workers who are organizationally below
your level deserve your special attention. Much of your work is
accomplished through cooperation from such persons, and you must
never get them in trouble with their own boss. You need to
establish operating ground rules with their superiors and keep
those superiors informed as necessary.

As an example, suppose that you are a quality engineer and
see a defective manufacturing process on the shop floor. You must
know whether you can stop this job by a conversation with the
machine operator, or whether you must see the department
supervisor. Perhaps it is neither of these and you must inform
your supervisor before anything can be stopped. The problem to
avoid is that the power of the engineer may cause a conscientious
operator to stop a job but later receive a reprimand from his
supervisor.

C1n: Higher Level Co-Workers

Conversation with co-workers who are organizationally above you requires careful reporting, and you should generally keep your own boss informed of such conversations. If you are a quality engineer, report conversations with the Production Manager to your boss, the Quality Manager. If such conversations are routine and frequent, then you should have an understanding with your superior as to how well he wishes to be kept informed.

C1o: CONVERSATION WITH SUPERIORS

As an engineer, you will likely have three different types of conversations with your superiors. You will have strictly technical conversations, conversations on plans and performance, and purely social conversations. Try to recognize which of the three situations exists during any particular conversation or portion of a conversation.

C1p: Technical Conversations

During technical conversations you must remain very objective. Clearly state what you know and do not know but be prepared to make educated estimates. Make certain that your superiors fully understand but without unnecessary detail, and most of all, make certain that you understand what they are saying to you.

C1q: Performance Conversations

When the conversation concerns plans, objectives, schedules, or performance you must listen between the lines and make sure that you understand what is in your boss' best interest as well as your own. After all, the best formula for success is to do everything possible to get your boss promoted. Although you may wish to improve a design on which you are working, your boss may have been given a production deadline. You and your boss must make a commitment to the best possible job within the allotted time frame. In most every instance, having an adequate product or service on time is more desirable than having a superior product too late.

C1r: Social Conversations

In social conversations with your boss, observe the rule of "don't talk business" unless your boss originates the conversation. Impress him with your knowledge and enthusiasm during business hours, not during church or cocktail parties.

C1s: CONVERSATIONS WITH SUBORDINATES

Sooner or later, every good engineer becomes a manager of some sort. You may be simply a group leader of other technical personnel or someday president of the company. In conversations with your subordinates, make sure that you have "calibrated" each of them. Calibrate their relative ability to listen, understand, and act.

Do not hesitate to request subordinates to repeat what they think you have said. This is particularly desirable with a newly acquired subordinate. Remember, as a manager you are responsible for what your subordinates do, not for what you told them to do. You will usually find that your subordinates occupy the full range from "he always understands" to "he never seems to understand."

The best rule is to be a defensive communicator much in the same manner as you would be a defensive driver. Assume that most everything you say will be misunderstood to some extent; therefore, take great pain to elaborate on the critical elements of the conversation. Do not tell your technician to put a larger fan on a power supply without telling him if there are any cost or time limitations on doing so. Otherwise, you may end up missing a critical schedule or cost target. Your cool power supply will have produced a "hot" boss or customer.

C1t: CONVERSATIONS WITH CLIENTS

Conversations with clients are everyday occurrences for sales and application engineers but are often rare for research, design, or manufacturing engineers. A course in salesmanship is beyond the scope of this communications manual, but it is the responsibility of every engineer within an organization to provide the client or customer with the best possible service or product at an affordable cost. A few simple rules on conversations with clients or customers are as follows:

- Place additional emphasis on listening rather than speaking. Make sure that you understand what he wants or to what he objects.

- Do not wash your dirty laundry. You may have a severe internal departmental problem, but your client won't solve it for you. Don't say something like "we designed the product on time, but our manufacturing department is always late." When speaking to the customer, you are speaking for your entire organization, not your department.

- Do not make promises that are beyond your control. Answer such requests from the client with "I'll get back to you on that next Tuesday."

- Most important of all, remember that your business or organization exists to satisfy its clients or customers. The customer is not always right, but he is always the customer.

C1u: REMEMBER

To be articulate is possibly the greatest engineering skill of all. The inarticulate lose credit for their own ideas and often hear them return from the mouths of others. Learn to speak by listening for what needs to be said.

CHAPTER C2
USING THE TELEPHONE

C2a: <u>ANALYZE FOR BETTER COMMUNICATIONS</u>

Your engineering activity will frequently involve use of the telephone. This chapter will discuss <u>when</u> to use it and <u>how</u> to use it.

C2b: <u>WHEN TO USE THE TELEPHONE</u>

Your telephone is a powerful tool and is particularly effective in the following situations:

- When distance, time, or cost precludes a face-to-face conversation.

- When the matter is routine, such as making a reservation or placing an order.

- When the topic is already well understood, and no personal interplay or persuasion is required.

However, there are certain other situations where the telephone <u>should be avoided</u> if at all possible:

- When discussing matters such as salary adjustments, promotions, reprimands, or terminations.

- When discussing highly technical matters involving complex equations or symbols.

- When a face-to-face visit is equally convenient.

- When a written document would better ensure proper action or a more thoughtful reply. You will find that some of your co-workers "forget" telephone promises.

C2c: <u>HOW TO USE THE TELEPHONE</u>

Your effective use of the telephone involves procedures, courtesy, clarity, and cost.

C2d: <u>Procedures</u>

Procedures can be summarized as follows:

- Do not speak any longer than necessary to conduct your business effectively. Allow time for a few courteous greetings but do not originate any "bull sessions." Others may need the phone line, or the person at the other end may be very busy and highly annoyed.

- Place your own calls whenever possible rather than go through a secretary. This not only speeds up the process but ensures that you, and not your secretary, will be on the line when the other party is called to the phone. No one appreciates being asked to "Hold for Mr. Jones."

- If you are disconnected on a long distance call, the calling party should reinstate the call, and the receiving party should have hung up.

- Do not answer a business phone with "hello." You should answer with your name, department, or phone number, whichever seems appropriate in a given situation.

- Verify your connection at least every 30 seconds. If you are doing most of the talking, ask for some comment that will verify that the other party is still connected. If you are doing most of the listening, say phrases like "right" or "okay" to let the speaker know that you are still there. Without such precautions you can talk for ten minutes to a disconnected line.

- Avoid making blind calls whenever possible. If you are calling an organization for the first time, try to find out the name of someone in that organization even if you know they are not in the department you want.

C2e: Courtesy

The following are a few rules of telephone courtesy:

- Always be polite with operators and secretaries, and you will find that your telephone conversations are considerably more effective.

- Offer to take messages for your co-workers even though you are not their secretary. However, be sure that you correctly record name, telephone number, date, and time.

- Remember that the calling, or originating, party has the responsibility to end the conversation. If you are the receiver of the call, give the caller every chance to graciously do so. If this fails, you may have to "catch a plane" or take an "urgent call" on another line.

- Do not have your secretary say that you are "not taking any calls." There are admittedly times when you should not be disturbed, but your secretary should say that you are "not in", or "on another line," or "have someone in your office." However, such instructions should be the exception rather than the rule, or you may find that important callers abandon their attempts to call you.

C2f: Clarity

The clarity of your communication on the telephone is much more vulnerable than in a face-to-face conversation. It is much easier to misunderstand something over the telephone because both parties must talk and listen without the benefit of facial expressions, movements, and gestures. That is why delicate matters should never be discussed over the telephone. Before placing a call, determine your objectives and make a list of points you wish to cover. You should also develop the habit of taking notes during every telephone conversation. This will help you ask questions and to remember what was said or promised.

C2g: Cost

The cost of a telephone call should always be considered, just like any other controllable expense. Station-to-station calls are less expensive than person-to-person calls, and WATS line calls are less expensive than station calls but are usually not free. Most WATS line services cost additional money beyond a monthly allocation.

Learn the features of the particular phone system in your organization or office. Many systems today are equipped with call forwarding, call transferring, and other very useful features that facilitate your daily activities.   Unfortunately, many engineers do not bother to learn the procedures and thus limit the effectiveness of such systems.

Another cost effective use of the telephone is to transmit brief written messages.   Extensive written transmissions would justify teletype or telecopying equipment, but occasional written messages can be effectively transmitted by dictating to a secretary or stenographer at the other end.   In most cases, this is more convenient and less expensive than sending a telegram.

C2h: <u>REMEMBER</u>

The telephone is a valuable tool, but analysis of when and how to use it provides for much more effective communications.   Use it when you should, get off of the line as soon as practical, and take extra precautions that your messages are understood.

CHAPTER C3

MEETINGS

C3a: <u>ANALYZE FOR BETTER COMMUNICATIONS</u>

As an engineer, you will spend a great deal of time in meetings, many of which you must organize and chair. You must carefully analyze what is expected of you as a participant, or as the chairman, and do your homework.

A meeting contains the following essential elements:

- Purpose
- Arrangements
- Chairman
- Participants
- Minutes

C3b: <u>PURPOSE</u>

Do not call a meeting unless you have a clear cut purpose and do not attend a meeting without at least attempting to find out its purpose. If the purpose cannot be simply stated in a few words, either on an agenda or at the beginning of the actual meeting, then a successful meeting is highly unlikely.

C3c: <u>ARRANGEMENTS</u>

Once the purpose for a meeting is established, then the arrangements become the next critical item for the meeting's success. Arrangements consist of the following elements:

- Who is to attend
- When to meet
- Where to meet
- Ground rules for the meeting
- Seating arrangements
- Comfort

<u>C3d</u>: <u>Who</u> <u>Is</u> <u>to</u> <u>Attend</u>

You will often hear the expression that "meetings are expensive" and will be tempted to invite as few people as possible. However, the productivity or economic value of a meeting is what you must consider, not its size. A small meeting of the wrong people accomplishes little or nothing. Of course a large meeting of the wrong people is even worse.

Here are a few simple guidelines on who should attend a meeting:

If the meeting requires participants to contribute to a decision or course of action, then invite only those whose position, experience, or talent will make a contribution.

If the meeting is to inform, then invite only those who need to be informed. If a large number of people need to be informed, then you have the choice of one large meeting or numerous small meetings held throughout the organization.

<u>C3e</u>: <u>When</u> <u>to</u> <u>Meet</u>

Hold your meeting at a time and date that assures the best possible attendance of fully prepared participants. You can accomplish this by always giving at least 24 hours notice, sending an agenda in advance, avoiding Friday or pre-holiday afternoons, and checking on the availability of key individuals.

Above all, avoid calling spur-of-the-moment meetings just to "kick things around." Competent people already have their day planned, and you will not get their best input at such meetings.

<u>C3f</u>: <u>Where</u> <u>to</u> <u>Meet</u>

The meeting should be held at a location that assures the best possible attendance of fully prepared participants. This means pick a location convenient to the majority, not yourself.

<u>C3g</u>: <u>Ground</u> <u>Rules</u> <u>for</u> <u>the</u> <u>Meeting</u>

The ground rules of the meeting must be consistent with the meeting's purpose. Parliamentary procedures under <u>Robert's</u> <u>Rules</u>

of Order are rarely, if ever, required in a technical meeting. However, three other categories of ground rules are possible:

- Authoritative
- Majority rule
- Consensus

The authoritative ground rules exist when everyone has the opportunity and obligation to speak, but the leader must decide. This situation nearly always exists when the leader is the organizational superior of all participants.

The majority rule is an acceptable format if all participants are of equal value to the meeting. Unfortunately, you will find that this is seldom the situation in a technical meeting since some participants will be clearly recognized as having greater or less knowledge of the subject under discussion.

The consensus type of ground rule is by far the most common. The leader makes sure that everyone is heard and then says "I believe that the general consensus is...". This approach automatically gives appropriate weight to the comments of all participants.

C3h: Seating Arrangements

Seating arrangements clearly affect the outcome of a meeting and therefore should also be consistent with the meeting's purpose.

If the meeting is a demonstration or presentation, then seat everyone so that they can see and hear.

If the meeting is to jointly solve problems, then seat everyone facing one another in circular fashion if possible.

If a decision will be made by the meeting's leader, then place that leader at a "head of the table" arrangement.

Seating arrangements are nothing more than common sense consistent with the purpose of the meeting itself.

C3i: Comfort

You must deliberately plan the relative comfort of a meeting and not leave this important factor to chance.

If a meeting is expected to last more than an hour, then provide coffee, comfortable chairs, and correct room temperature.

If a meeting is to last more than three hours, then provide lunch or a lunch break.

If you wish to encourage a less than one hour meeting, then use negative comfort factors such as no coffee, or starting close to lunch or quitting time.

C3j: THE CHAIRMAN

If you are called upon to chair a meeting, you have the following responsibilities:

Your behavior will determine the group's behavior. In other words, you will get what you deserve. Carelessness will breed carelessness; boredom will breed boredom; arrogance will breed arrogance.

You must ensure proper attention to agenda and time limitations. Your ability to start a meeting on time, keep it on the track, and end on time is a key measure of individual ability and will greatly affect the success of your engineering career. Your best vehicle is to distribute an agenda in advance that states a closing time as well as a starting time. To aid starting on time, place important or controversial items first or be prepared to dispose of minor items without full attendance.

You must facilitate, clarify, and close discussion on each topic. Facilitate means to enlist participation commensurate with ability; clarify means to resolve conflict or misunderstanding by asking participants to elaborate and look for areas of agreement: and close the discussion means deciding that a decision can be reached, or that unavailable information, persons, or time preclude further discussion.

C3k: THE PARTICIPANT

You are more likely to be a participant rather than chairman in the early years of your engineering career. You must remember that the leader is not totally responsible for a meeting's success or failure and that you have the following responsibilities as a participant:

Prepare for the meeting. The amount of your preparation depends upon the amount of advance notice and any action items that you may have been assigned from the previous meeting. As a very minimum, however, set aside a quiet period immediately prior to the meeting and reflect on what is likely to be discussed. Above all, do not postpone your action items until the last minute and then rush to the meeting with your mind cluttered with detail.

Arrive on time. It is not fashionable to be late. If you arrive at a meeting late, others will assume that your input is proportional to your promptness.

You are obligated to speak when appropriate and listen when appropriate. Do not be guilty of dominating the discussion, carrying on side conversations, telling irrelevant "war stories", saying little or nothing, or having a closed mind.

C3l: THE MINUTES

In most technical meetings, the chairman himself should write the minutes. At the very least, he should review and initial the minutes prior to their issue.

C3m: When Are Minutes Required

You should use the general rule of thumb that any meeting worth having deserves minutes. However, minutes are a necessity under the following situations:

- Additional meetings are required, and minutes become the agenda for the next meeting.
- The group is responsible to a higher authority which requires regular reports.
- Decisions need formal documentation.
- Action items need assigning.

<u>C3n</u>: <u>Format</u> <u>of</u> <u>the</u> <u>Minutes</u>

An informal format is satisfactory for the minutes of a technical meeting. Such minutes need not be officially read and approved, and their purpose is to facilitate further activity.

Even though informal, the minutes should include the following items:

- The meeting's subject, date, time, and location.
- The names of those attending the meeting.
- Itemization of all decisions that were made at the meeting.
- Names and completion dates for all action items assigned.
- The time, date, and tentative agenda for the next meeting.

<u>C3o</u>: <u>REMEMBER</u>

Do not hold or attend a meeting without knowing its purpose and recognizing that the arrangements, as well as choice of chairman and participants, are key factors to a meeting's success.

CHAPTER C4
ORAL TECHNICAL PRESENTATIONS

C4a: ANALYZE FOR BETTER COMMUNICATIONS

As an engineer you will be called upon to make oral presentations throughout your career. To achieve satisfactory results, these presentations or briefings must be properly "engineered". This chapter does not replace a course in public speaking but will provide the fundamental guidelines that give confidence at the lectern.

The engineering of a good presentation needs to proceed in the following logical manner:

- Establish objectives
- Analyze the audience
- Organize the material
- Rehearse the presentation

C4b: ESTABLISH OBJECTIVES

Speeches are often categorized as either explanatory or persuasive, but an effective technical presentation must really be both. A sales presentation to a prospective customer must still explain the product, and a classroom lecture must persuade the student to study and understand certain material.

The two fundamental questions are "Why am I making this presentation?" and "What result do I expect?" It is important that you answer these questions in realistic terms confined to the single presentation at hand. A long-range objective might be to design and produce a new electric car, but the particular presentation in question may be to provide a monthly update on technical alternatives being pursued and to obtain approval from engineering, marketing, and financial management to prototype one or more of those alternatives.

C4c: ANALYZE THE AUDIENCE

To be effective, a presentation must "fit" the audience. Find out the names and functions of those who will attend. This will give the size of the audience and an indication of their knowledge, interest, and influence in the subject at hand. Learn the names and faces of the key individuals who can help you accomplish your desired results. Nothing is more effective than answering a question from the floor with "Yes, Mr. Jones, that is one of our main considerations."

There are five key stages of listening, and you must be continuously aware of the stages and their transitions as they pertain to the influential members of the audience:

- Awareness (Why am I here?)
- Willingness to Listen (Do I want or need to hear this?)
- Understanding (I understand, but do not necessarily agree with what is being said.)
- Acceptance (I agree with what is being said or will not accept the consequences of disagreeing.)
- Behavior (I will take the necessary action recommended in this presentation.)

Guiding the key members of an audience through these stages requires an understanding of both their knowledge of and interest in the subject presented. It is important to remember that a "communication" is defined as the message received and not the message transmitted.

C4d: ORGANIZE THE MATERIAL

This involves an outline of what is to be said, selection of resource and reference material, and an orderly arrangement for presentation. This step of preparing and organizing the message is obviously the essential phase of a presentation and must be done well in order for the overall presentation to be effective. Nothing will cover up a poorly organized message. However, you must recognize that the best of messages can be ruined by the speaker's failure to consider his audience, his image, or his delivery.

C4e: Outline

The preliminary plan must include the previously established objectives and audience information and then list the main ideas to be covered. No more than five or six main ideas should be attempted in a single presentation. If your intentions are more complex, then arrange two or more presentations.

C4f: Select

The selection of material is heavily dependent upon the audience and the necessity to guide key members through the "five stages of listening." Remember that messages can convey ideas, conceal ideas, or conceal the absence of ideas.

Since the engineer is seldom called upon to make a purely technical presentation, sufficient material must be gathered to satisfy economic, marketing, and production considerations. Even if only engineering department personnel and management are present, you must give your Engineering Manager insight to these other areas.

It is most important that information in the non-technical areas be clearly defined as to whether it is the engineer's estimate, not "guess," or whether it is an estimate supported or made by a responsible individual in another department. Naturally, the latter situation is much preferred if preparation time permits.

C4g: Arrange

Organizing the collected material for presentation is the critical step. You should follow the classic sequence of "Introduction," "Body," and "Conclusion." If the speaker is unkown to his audience, his credentials and relationship to the technical activity need to be described. This is best done by a person introducing the speaker or by a written statement on the program or agenda.

C4h: The Introduction

The introduction must answer the listener's first two questions, "Why am I here?" and "Do I want or need to hear this?" Unfortunately, there are often members of the audience who cannot properly answer these questions, regardless of the speaker's effort.

Most technical presentations are best served by having the introduction be a direct statement of the subject and why it is important. Indirect introductions using a related story, or statistics, are sometimes appropriate but usually require more skill for delivery.

The length of the introduction can vary between 5%-15% of the total presentation. However, a long introduction should never be used to disguise inadequate material for the body. It is much better to shorten the entire presentation.

C4i: The Body

The body of the message must lead the audience from the known to the unknown and should be organized around <u>Time</u> or <u>Ideas</u>. For example, during a project definition phase, a progress report would be organized around <u>Ideas</u>. During the development phase it would be organized around <u>Time</u>.

The body of the presentation must cover the main points listed in the plan and will be around 80% of the total presentation.

C4j: The Conclusion

The conclusion should be used to restate the message. This is the place to "tell them what you've told them," solicit acceptance of the message, and request appropriate action. It is best not to introduce any new ideas in the conclusion.

C4k: <u>REHEARSE</u>

The purpose of practice is not simply to become familiar with the message and lecture notes but to refine the delivery, image, and integration of visual aids.

C41: Delivery

The use of a tape recorder is the best means of practicing for a refined delivery. There are four essential elements of delivery:

- Grammar (Sentence structure and pronunciation)
- Voice (Tone, flow, volume, and enunciation)
- Timing (When to pause, when to emphasize)
- Enthusiasm (Do you believe what you are saying?)

These are all _audible_ characteristics, and you can make dramatic improvements by utilizing only one playback of the first rehearsal.

C4m: The Image

Studying and improving the image projected by the speaker are somewhat more difficult without the obvious benefit of a _video_ recording. However, almost anyone can watch your first rehearsal and provide extremely valuable feedback. Enlist comments on the following:

C4n: Personal Appearance

Be sure that you are appropriately dressed and well groomed. When in doubt, it is always better to be slightly over-dressed.

C4o: Facial Expressions

Facial expressions are one of the most visible forms of nonverbal communication. Spoken words can be over intensified, de-intensified, neutralized, or masked. Said more simply, you can act more sad than you are, less sad than you are, convey no emotion at all, or act happy when actually sad.

There is a place for nearly every type of facial expression during a normal presentation. You should not be afraid to smile, or frown, or to look the audience in the eye.

C4p: Body Movements

Avoid excessive pacing and poor posture, and try to appear relaxed and under self-control. Hand gestures must be used judiciously since overuse of gestures is much worse than no gestures at all.

C4q: <u>VISUAL AID</u>

The visual aid is a valuable "cue card" and has become a standard component of most technical presentations. Items such as flip-charts, overhead transparencies, or 35 mm slides add greatly to the integrity of the message received by the audience. Many people are readers, not listeners, and visual aids provide the necessary emphasis.

The danger of visual aids is their overuse to a degree that they become a visual distraction. You should not feel compelled to have something on the screen at all times, but do not turn the projector off and on. Use a blank slide or page to darken the screen whenever appropriate.

Above all, each visual aid, regardless of type, must be readable from the most distant corner of the audience. An overhead transparency or slide should only be a few words or numbers and in large print. Do not read or expect your audience to read a copy of a normal type-written page.

Rehearsal of the presentation with all visual aids employed is absolutely necessary. If you cannot rehearse in the room to be used, then be sure that all necessary equipment such as screens, projectors, and PA systems will be available and usable for the actual presentation. See Chapter D8 for a further discussion of visual aids.

C4r: <u>HANDOUTS</u>

The distribution of written material to the audience has serious implications on the delivery and other visual aids. If you distribute handouts at the beginning of the presentation, most of the audience will read rather than listen, and a presentation becomes little more than a supervised library session. If the handouts are distributed one or more days before the presentation, then the presentation becomes more of an interrogation of the speaker.

In most instances, it is strongly recommended that all handouts be made _after_ the verbal presentation and thus used as a reminder of necessary action by key members of the audience.

C4s: <u>REMEMBER</u>

As an engineer, your ability to prepare and deliver an effective oral presentation is fundamental to your success. In the early part of your career, such presentations must be carefully engineered because they convey your "apparent" knowledge as a practicing professional. If you cannot make an effective oral presentation, you will always be depending upon <u>someone else</u> to sell your ideas and talent.

CHAPTER C5
ERROR MESSAGES - ORAL PRESENTATIONS

C5a: <u>ANALYZE FOR BETTER COMMUNICATIONS</u>

A critical analysis of every presentation will give you the important feedback required for improvement.

C5b: <u>ERROR MESSAGES</u>

Example: Error Message, AK5a, means that the audience's knowledge of the subject was inadequate because material was too technical.

<u>Audience Related</u> --(refers to the audience's)

AP • Awareness of speaker's purpose.

AL • Willingness to listen.

AK • Knowledge of subject.

AI • Interest in subject.

AU • Understanding of message.

AA • Acceptance of message.

<u>Image Related</u> --(refers to the speaker's)

IA • Personal appearance.

IF • Facial expressions.

IB • Body movements.

IG • Gestures.

IP • Posture.

IC • Poise (Confidence).

<u>Delivery Related</u> --(refers to the speaker's)

DG • Grammar.

DT • Voice tone.

DF • Voice flow.

DV • Voice volume.

DS • Timing (sense of timing).

DE • Enthusiasm.

DP • Emphasis (Proper Emphasis).

<u>Message Related</u> --(refers to the message's)

MP • Preparation.

MO • Organization.

MI • Introduction length/content.

MB • Body length/content.

MC • Conclusion length/content.

Visual Aids --(refers to the aids')

VT • Type.

VQ • Quality.

VN • Quantity (number).

C5c: ADJECTIVE CODE

0   Superfluous

1   Marginal

2   Distracting

3   Awkward

4   Uninteresting

5   Inadequate

6   Inappropriate

7   Improper

8   Excessive

9   Offensive

C5d: CORRECTIVE ACTIONS CODES

Audience-Related - A Prefix

a.   Material too technical

b.   Material too elementary

c.   Material not culturally suitable

d.   Speaker was not properly introduced

e.   Speaker's credentials not adequate

Image-Related - I Prefix

a.   Clothing too formal

b.   Clothing too informal

c.   Grooming improper

d.   Too stern

e.   Too flippant

f.   Too happy

g.   Excessive pacing

h. Turned back to audience

i. Unnatural stance

j. Too much pointing

k. Untimely gestures

l. Leaning against lectern or wall

m. Slumped shoulders

n. Did not look at audience

o. Too arrogant

p. Too apologetic

q. Hands in pockets

Delivery-Related - D Prefix

a. Language too formal

b. Language too informal

c. Too many "big" words

d. Awkward sentence structures

e. Too many grammatical or pronunciation errors

f. Nervous voice pitch

g. Too many "and-uh's" and "ok's".

h. Spoke too rapidly

i. Spoke too slowly

j. Insufficient pauses

k. Spoke too loudly

l. Spoke too softly

m. Appeared hyper

n. Appeared bored

o. Key points not stressed

p. Too many points over-stressed

q. Poor enunciation

r. Overuse of notes

Message - Related - M Prefix

a. Speaker's subject knowledge questionable

b. Audience questions not answered adequately

c. Sequence illogical

d. Too many sub-topics

e. Insufficient detail

f.   Excessive detail

g.   Too much irrelevant material

h.   Too much exaggeration

i.   Illogical arguments

j.   Points not obvious

k.   Audience left "dangling"

l.   Statements untrue

<u>Visual</u> <u>Aid</u> <u>Related</u> - <u>V</u> <u>Prefix</u>

a.   Unreadable - too small

b.   Too cluttered

c.   Not focused properly

d.   Not in view of whole audience

e.   Poor artwork

f.   Poor illumination

g.   Unrelated to verbal message

h.   Out of sequence with verbal message

i.   Dominated the presentation

j.   Too elaborate or sophisticated

k.   Too crude

CHAPTER C6

REFERENCES FOR ORAL COMMUNICATIONS

Lee, Irving J. How to Talk with People. New York: Harper & Brothers, 1952.

Morrisey, George L. Effective Business and Technical Presentations. 2nd ed. Reading, MA: Addison-Wesley, 1975.

Oliver, Robert T. Conversation-The Development and Expression of Personality. Springfield, IL: Charles C. Thomas, 1961.

Palmer, Barbara C. and Kenneth R. The Successful Meeting Master Guide, Englewood Cliffs, NJ: Prentice-Hall, 1983.

Tropman, John E. Effective Meetings. Beverly Hills, CA: Sage Publications, 1980.

Vohs, John L., and G.P. Mohrmann. Audiences Messages Speakers. New York: Harcourt Brace Jovanovich, 1975.

Persuasion and the Role of Visual Presentation Support, a study by the University of Minnesota and the 3M Corporation, Minneapolis.

Six Secrets to Holding a Good Meeting, Publication #78-6970-0944-3, Minneapolis: Audio Visual Division, 3M Corp.

NOTES

SECTION D
GRAPHIC COMMUNICATIONS

More often than not your technical presentations, both written and oral, will depend upon more than words to get your message across. Drawings, sketches, photographs, tables, charts, graphs, and other visuals are used to reinforce, explain, and emphasize your ideas. It is extremely important that these visuals be professional in every way. An otherwise excellent report can be severely damaged by carelessly drawn sketches, sloppy graphs and charts, poor photographs, or lettering that looks bad. Chapters D1 through D8 will acquaint you with the principles of making good visuals.

Carefully review Section A on analyzing your objectives and audience since these steps are most important in the proper selection and use of a graphic communication.

CHAPTER D1

LETTERING AND PRINTING

D1a: ANALYZE FOR BETTER COMMUNICATIONS

All sketches, drawings, tables, and charts have many of the details communicated through words and symbols, and this lettering should be more uniform and easier to read than script.

Whether freehand lettering your engineering sketches or using automated lettering devices for a more formal output, you should always analyze your drawing to determine the type, size, and amount of lettering needed. As in every other communication situation, review the needs and objectives of your readers and provide the proper amount of information to accomplish your goals.

D1b: FREE-HAND LETTERING

Every engineering sketch, no matter how simple, includes a certain amount of writing. To ensure legibility, clear free-hand lettering is expected as a matter of course from all who use the graphic language as a means of communication.

D1c: Single-stroke Lettering

The most commonly used style of lettering in engineering graphics is called single-stroke Gothic. Designed for legibility and easy execution, the strokes may be either vertical or inclined (see Fig. D1-1). As an engineer you will probably develop your own personal style of lettering. However, to maintain clarity in your writing, you should become familiar with the characteristics of the letters and numerals used in single-stroke Gothic, specifically the shape and general proportions of each one of them.

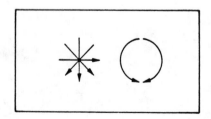

FIGURE D1-1

Basically, all letters and numerals are composed of straight lines, circles and ellipses, or combinations of all three (see Fig. D1-2). The orderly succession of strokes is very important in the execution of good free-hand lettering. These strokes should be made from top to bottom and left to right, so that you are always pulling on the pencil.

You should also remember the general proportions of each letter, from the narrowest, I, to the widest, W. Except for these two, all uppercase letters should be almost as wide as they are high.

D1d: Use of Guidelines

Guidelines are necessary to attain uniformity in free-hand lettering and to obtain legible, well-proportioned letters and numerals. Draw these guidelines very lightly to avoid detracting from the lettering itself. For engineering sketching you can use the lines of grid paper as guidelines, thereby avoiding the need to draw them. A square, 1/8" grid will usually give satisfactory results (see Fig. D1-3). Always use some sort of guidelines when lettering free-hand; only poor lettering will result if you do not.

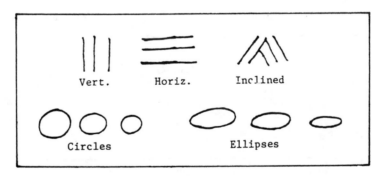

FIGURE D1-2

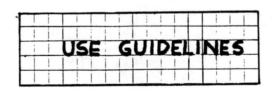

FIGURE D1-3

117

D1e: Spacing for Composition

To be able to attain better legibility when lettering, you must not only know the proper form of each letter but also how to space letters in words and words in sentences. Improper spacing of letters and words will cause sentences to be very difficult to read, no matter how well formed each letter is.

The person reading your drawing will usually see groups of from 2 to 10 words at one time, and then will move to the next group. Individual letters will not be noticed unless they are poorly formed. Remember that each word must appear as a unit.

Letters in a word cannot be placed at equal distances from each other. They should be arranged so that the area between adjacent letters appears to be the same. It is the shape of each letter and the shape of the letter preceding it that control the distance between them (see Fig. D1-4).

Spacing of words follows the same guidelines. Figure D1-5 is an example of good word composition.

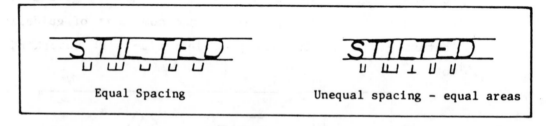

Equal Spacing	Unequal spacing - equal areas

FIGURE D1-4

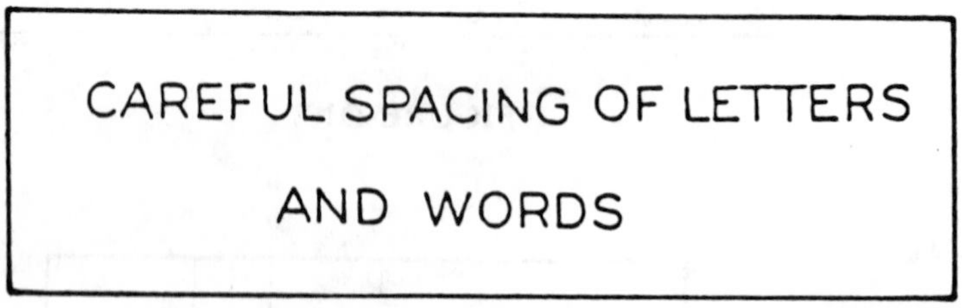

CAREFUL SPACING OF LETTERS

AND WORDS

FIGURE D1-5

D1f: <u>AUTOMATED LETTERING</u>

If your drawings are to be reproduced, you would probably like to improve your lettering through automated techniques. Remember, however, that resorting to mechanical devices or transfers will slow down the pace of your lettering.

The use of computer-generated letters is presently becoming very common in engineering practice, especially due to the availability of relatively inexpensive computer-aided design and drafting systems. Section D5 in this manual will cover this area in more detail.

D1g: <u>Basic Tools and Practices</u>

Many types of lettering devices are available in the market to help with letter construction. With most of them, however, you still need to pay attention to the problem of spacing between letters and words.

Lettering aids used by engineers can generally be divided into template-type and transfer-type. Simple lettering templates are built of punched-out alphanumeric shapes which can be duplicated by outlining the contours of the holes with lettering pens or pencils (see Fig. D1-6). Another type of lettering aid makes use of a grooved template. A guide pin follows the contour of the letters grooved on the template, while a tracer reproduces them on the paper (see Fig. D1-7). The templates are always placed against a fixed straightedge, which serves as a base guideline, and are moved sideways to position the letters.

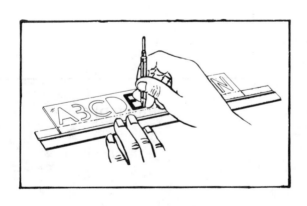

FIGURE D1-6

FIGURE D1-7

119

The transfer-type lettering aid most commonly used is the dry-transfer. It is composed of preprinted letters and symbols in a variety of sizes and shapes attached to a clear plastic or translucent paper with a wax-like adhesive on the back (see Fig. D1-8). To obtain good results you still use guidelines and rub over the letter with a smooth stylus.

With all these lettering devices you should follow the general rules of spacing for composition outlined in section D1e.

D1h: GENERAL PRINTING TERMINOLOGY

Several terms used in the printing industry and related to text output have become of common use, especially with the advent of computer-generated text software. When preparing an oral or written presentation, you need to be generally familiar with this terminology in order to analyze the styles of text available and find those best suited for your particular application.

The most basic group of images in graphic design is the Latin alphabet, which can be reproduced in many different "type-faces." Remember that the design of any character is based only on variations of line thickness, slant, and "serifs", which are short crosslines placed at the end of unconnected strokes in the letters (see Fig. D1-9).

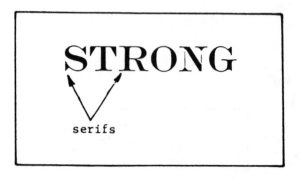

FIGURE D1-9

FIGURE D1-8

The most commonly used type-faces are these:

GOTHIC - All elementary strokes have even width (see Fig. D1-10). Notice that no serifs are present in Gothic-style letters. From this style originated the modern single-stroke engineering letters discussed in section D1c. They are easy to reproduce and highly legible. In the general printing industry, Gothic letters are very effectively used for titles and headings.

ROMAN - All elementary strokes consist of heavy and light lines (see Fig. D-11). Serifs appear in all Roman-style letters. Most books and magazines are set in Roman typeface.

Several novelty-styles are also constantly being created and can be successfully used in place of the more traditional styles described above.

ENGINEERING

FIGURE D1-10

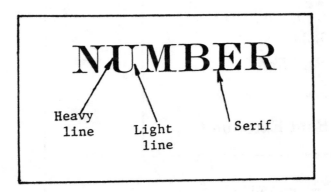

FIGURE D1-11

Each of the general typefaces can show variations by simple manipulation of the basic design (see Fig. D-12). The word "font" represents the total collection of letters and symbols in each variation for one specific size (see Fig. D-13).

Always remember, especially when preparing text-visuals for oral or written presentations, to choose the most appropriate letter-style. Section D1k will elaborate further on this topic.

D1i: TITLES, LABELS, AND TEXT SLIDES

Every drawing you prepare will have a descriptive title and labels, giving information on its contents. The wording of the title and labels is very important, but its letter-style, size, and arrangement on the paper or slide may be the dominant factor in attracting attention to the message.

D1j: Title and Label Composition

Three major factors control the composition of a title or label in a drawing:

1. Type-style
2. Type-size
3. Positioning on the paper.

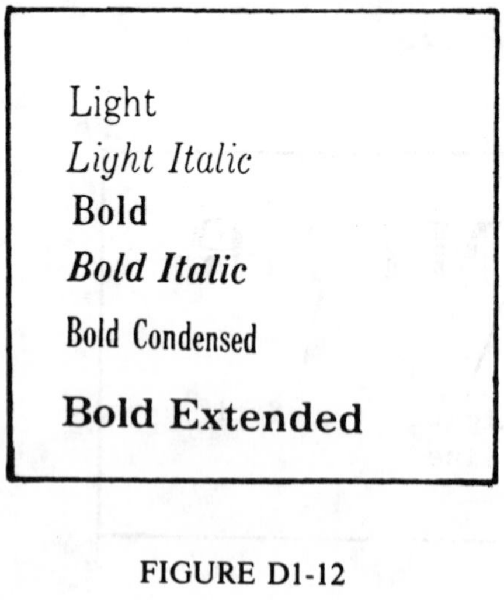

Light
Light Italic
Bold
Bold Italic
Bold Condensed
Bold Extended

FIGURE D1-12

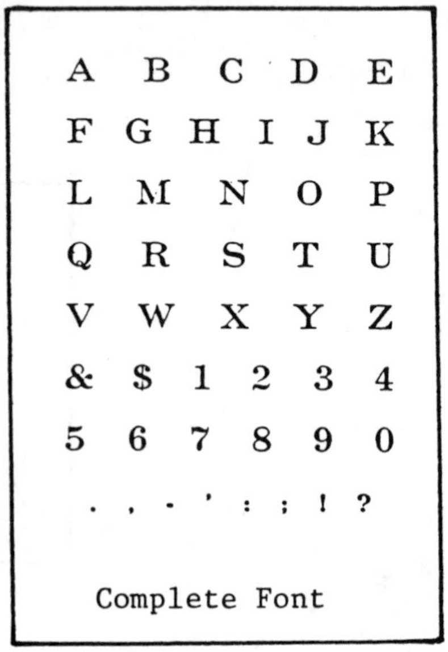

A B C D E
F G H I J K
L M N O P
Q R S T U
V W X Y Z
& $ 1 2 3 4
5 6 7 8 9 0
. , - ' : ; ! ?

Complete Font

FIGURE D1-13

The standard type-style for titles and labels is the Gothic-style, traditional for engineering lettering. However, if working with computer-generated text, you may be tempted to use the large variety of styles available to you. Be cautious - the simple type-styles are still the most appropriate for your drawing output.

The size used should be large enough to stand out in competition with other elements in the drawing. When deciding on the size for a title or label, keep in mind the following factors:

- <u>Weight of other elements in the display</u>. The title or label should draw the attention of the reader.

- <u>Space available</u>. Longer titles may force the use of a letter size that is too small. This problem can be circumvented by breaking the title or label into major and minor groups.

If working on a written report, select a size that will not be smaller than 1/16 inch when reduced.

You may position titles on the top or bottom of your drawing. They may be centered or left- and right-justified. Labels are usually placed close to the drawing area to which they refer so that they will not detract from the drawing itself.

<u>D1k</u>: <u>Overhead Transparencies and Slides - General Guidelines</u>

When preparing text transparencies and slides, try to follow these general guidelines and refer to Chapter D8 for details:

<u>Do not put too much text in each slide or transparency</u>. Each line of text should represent a complete thought. Do not use more than six lines and a maximum of five words per line in a slide. Transparencies can accommodate a little more text. Overcrowded slides and transparencies result in letters that are too small to be read comfortably at a distance. Titles should be short and preferably contained in one line.

Select simple type-styles. For slides and transparencies it is safest to restrict yourself to simple styles, like the modern Gothic and Roman. To create a sense of unity throughout your presentation, avoid mixing type-styles. Use a larger type-size for titles, and keep the same title size within the total presentation.

Use a balanced composition. Balance in the text layout gives the audience a sense of equilibrium (see Fig. D1-14). Once a layout form is chosen use it consistently throughout the presentation.

D11: REMEMBER

The most important rule to follow when using text is clarity. Try to develop a lettering style that is consistently clear, and avoid crowding text on transparencies or slides when preparing oral presentations. If you put too much text on a slide or transparency, your audience may not read any of it.

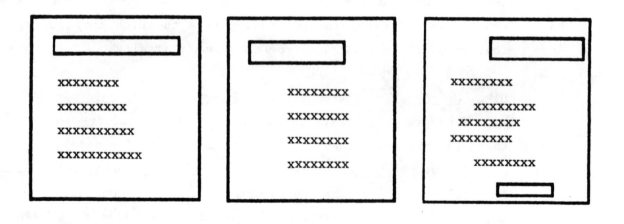

Formal Balance.   Informal balance.   No balance.

FIGURE D1-14

CHAPTER D2
SKETCHING

D2a: <u>ANALYZE FOR BETTER COMMUNICATIONS</u>

The ability to communicate an idea or design to others is very important to an engineer, and sketching plays an essential role in this communication process. A good sketch is also very useful to clarify your own ideas when working on a design project. As in all communication situations, review the needs of your audience as you prepare technical sketches and apply the analysis procedures outlined in section A.

D2b: <u>IMPORTANCE OF HAND SKETCHES IN ENGINEERING PRACTICE</u>

It is difficult to describe a complicated technical object in words. Furthermore, people interpret words differently. A sketch eliminates the need for involved word descriptions and gives a more accurate representation of an object. It plays, therefore, a very important role in any written or oral technical presentation. You can also use free-hand sketches to help visualize a physical problem and organize your thoughts. You should learn to use sketching instinctively, to "think through" your ideas as they gradually evolve into workable realities. Sketching is nothing more than an informal type of drawing. A good free-hand sketch must be simple and easy to understand. You should be able to prepare it quickly.

D2c: <u>LINE SKETCHING TECHNIQUES</u>

Free-hand sketching is based on simple line drawing techniques. The only materials needed are pencil, paper, and eraser. Sketches and diagrams can be easily prepared when the appropriate techniques are followed.

D2d: <u>Sketching Straight and Curved Lines</u>

All free-hand sketching involves the construction of straight and curved lines to define the shape of an object. To draw a straight line, hold the pencil at least one inch from its point, as shown in figure D2-1. Mark or imagine the distance between two points where you expect the line to start or finish. Beginning at point a, move the pencil towards point b with either short overlapping strokes or one single sweeping motion while fixing your eyes on the ending point b (see Fig. D2-2). Do not use the wrist or elbow since this will tend to bend the line. Move your hand, pencil, and arm at the same time in the direction away from your body. If not successful the first time, keep trying with very light strokes until you have found the correct position for the lines. Once this is done, darken those you wish to preserve by tracing over them.

You may find that it is easier for you to draw horizontal rather than vertical lines. Change the position of the paper to suit your preference. For many people a slight angle is preferred for free-hand strokes. If, however, you choose to draw vertical lines, the stroke direction should be from top to bottom. Parallel lines are made by keeping the eye on the position of the first line drawn as you sketch the second.

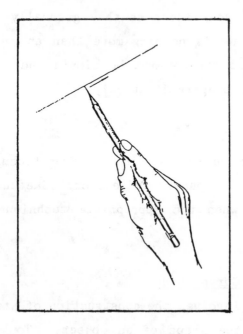

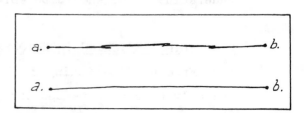

FIGURE D2-2

FIGURE D2-1

Circles or symmetric curves are best drawn by marking a few key points as guides. Figure D2-3 shows the procedures for sketching circles and ellipses, starting from the shape of a square or a rhombus and dividing them into quadrants.

You may find your ability to sketch greatly improved by using a grid paper. This will give you a reference for straight line construction and will help you establish the basic points needed for the development of curved lines.

D2e: Proportions

The ability to present details and overall mass of an object in its correct proportion is essential to free-hand sketching. You should always start your sketch by lightly "boxing-in" the overall size of the object, which means finding the correct relationship among its overall dimensions. Next, proportionally locate the major masses or areas with respect to this overall size. Details should be left for the end and worked only after the major

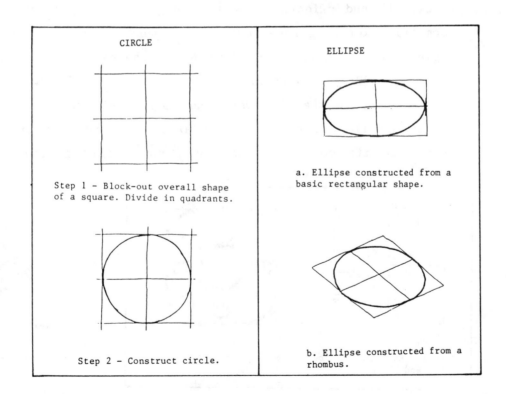

FIGURE D2-3

shapes and sizes have been established. Figure D2-4 shows an example of this sequence in a 2-D sketch. The same technique applies to 3-D sketches and will be further discussed in section D2g.

D2f: <u>SKETCHING IN THREE DIMENSIONS</u>

Three-dimensional sketches, also called pictorial sketches, are used to show an object in a manner that approximates the way you see the object when actually looking at it. Pictorial sketches usually provide overall appearance and general relationships rather than detailed information.

D2g: <u>Basic Methods</u>

Several methods are available for pictorial representation of objects. The two most commonly used in engineering practice are called "isometric" and "oblique."

The isometric pictorial has the advantage of being simple, practical, and effective for most engineering applications. In isometric sketching three mutually perpendicular axes, with equal angles between them, are used to define the direction of the lines (see Fig. D2-5). The corner of an object will always be its closest part to the observer, and horizontal and vertical lines remain parallel to the isometric axes. In order to present the object in its correct proportions, you should follow the same

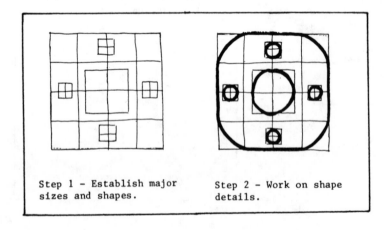

Step 1 - Establish major sizes and shapes.

Step 2 - Work on shape details.

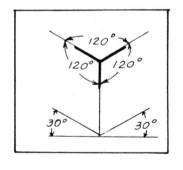

FIGURE D2-5

FIGURE D2-4

procedures outlined in section D2e. Figure D2-6 demonstrates these procedures, which in 3-D start with the outline of a box having the overall dimensions of the object, from which details of the contour are cut out. Without erasing any construction lines, which help in the general understanding of the shape of the object, darken the lines indicating the final contour.

If circular features are present, they will appear as ellipses on the isometric sketch. In this case you may prefer to use an oblique pictorial, which contains an undistorted plane parallel to the paper. Figure D2-7 shows the same object sketched in oblique and isometric forms. Notice how the circles and curves remain undistorted in the oblique sketch.

In choosing between the two pictorial systems to represent an object, keep in mind that oblique views generally show more

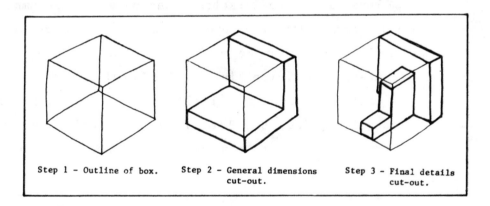

Step 1 - Outline of box.   Step 2 - General dimensions   Step 3 - Final details
                                       cut-out.   cut-out.

FIGURE D2-6

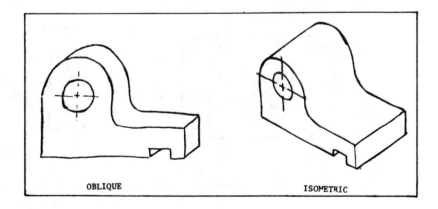

OBLIQUE   ISOMETRIC

FIGURE D2-7

distortion than isometric views.   They   have,   however,   the
advantage mentioned above of being capable of representing circles
and curves in their true shapes.   Both pictorial systems can be
executed easily and rapidly.

Besides choosing which of the sytems best suits your needs,
you must also decide on the most appropriate orientation for the
object.   Figure D2-8 shows pictorial views of an object in two
orientations.   Notice how (a) communicates the details of the
contour better than (b).   Before starting a pictorial sketch study
the object and establish the orientation that best communicates
its shape.   Sometimes it may help to include sketches in two or
more orientations.

D2h: Use of Grid Paper

Three-dimensional sketches can be easily accomplished by the
use of appropriate grid paper.   The grid lines are usually printed
in light green, orange, or blue so that they are visible but do
not interfere with the lines on the final sketch.

While oblique pictorials can be sketched on standard
rectangular grids (see Fig. D2-9), an easier representation of
isometric pictorials can be accomplished by the use of an

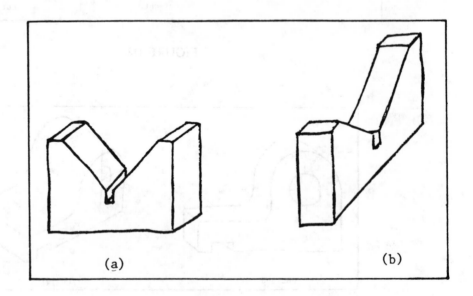

(a)   (b)

FIGURE D2-8

"isometric" grid (see Fig. D2-10). Notice that the direction of the isometric axes is established on the paper, allowing for the fast execution of lines on the drawing.

Use grid paper whenever possible to facilitate the construction of your sketches, both two- and three-dimensional.

D2i: Shading and Emphasis

Many three-dimensional sketches can be greatly improved by adding some simple shading to bring out the shape quickly and clearly. For most engineering purposes this is not necessary, but there will be times when this effect can and should be used to improve appearance.

To indicate differences in surface shapes follow these rules:

1. Flat surfaces - Shade with parallel lines, which can be vertical or parallel to the pictorial axes (see Fig. D2-11).

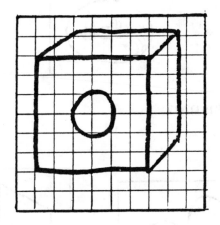

FIGURE D2-9

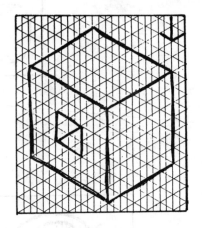

FIGURE D2-10

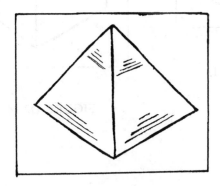

FIGURE D2-11

2. <u>Curved surfaces</u> - Shade by the use of lines parallel to the contour, as shown in figure D2-12. If hollow cylinders are present, shading should be placed on the opposite side (see Fig. D2-13).

3. <u>Surface transition</u> - To represent where curved and flat surfaces join, parallel lines can be used as indicated in figure D2-14. If the intersection of two flat surfaces does not form a sharp angle, lines can be sketched to indicate the curved transition, as shown in figure D2-15.

Lines can also be added for emphasis in engineering sketches, although this is never done in formal drawings. For example, if you wish to show that an object is moving in a particular direction, sketch a few lines on its back indicating the path being followed. To show the flow of fluids, draw a few straight

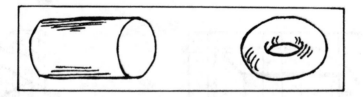

FIGURE D2-12

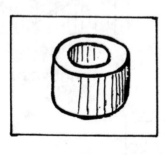

FIGURE D2-13

FIGURE D2-14

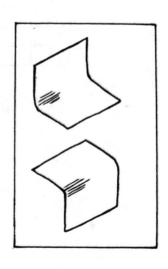

FIGURE D2-15

132

lines in the direction of movement.   Figure D2-16 illustrates
these examples.   Arrows can be very helpful in understanding the
operation of certain devices, as shown in figure D2-17.   To
further clarify this operation, multiple positions of an object
can be sketched as indicated in figure D2-18.

D2j: REMEMBER

Free-hand sketches will help you describe your ideas clearly and
accurately.   They will also serve as an aid in organizing your thoughts.
Use appropriate grids whenever possible to facilitate the execution
process.   Emphasis lines will improve your sketch by quickly bringing
out the object's shape.   Keep in mind your goals and the objectives of
your readers so that you include just enough detail.   Avoid "turning-
off" or confusing your readers with excessive information.

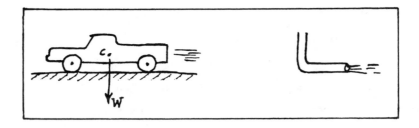

FIGURE D2-16

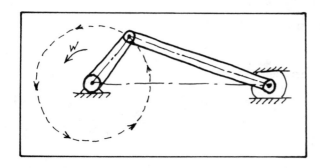

FIGURE D2-17

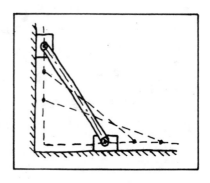

FIGURE D2-18

CHAPTER D3
SCALED DRAWINGS

D3a: ANALYZE FOR BETTER COMMUNICATIONS

Scaled drawings are an essential communication tool for the production of physical things. In the engineering profession, you are likely to use formal or informal scaled drawings daily. Whether preparing or simply interpreting them, follow the four-step analysis process outlined in section A of this manual.

D3b: IMPORTANCE OF SCALED DRAWINGS IN ENGINEERING PRACTICE

Scaled drawings are used to give accurate details on any new or changed design so that manufacture to specific requirements can be easily performed. Dimensions and notes are as essential to the description of a part as is its shape. They provide specifications and sizes of different features, and can be used both in formal working drawings and in informal design drawings for a complete communication of ideas.

Use rough sketches to organize your thinking before beginning a scaled drawing. Errors may be costly at this stage of a project.

D3c: SELECTING AN APPROPRIATE SCALE

Complete indication of an object's size and the exact relationship of its different parts requires an appropriate scale for the drawing. It is often impossible to prepare a full-size drawing, where actual dimensions are used. Objects may be too large as, for example, an airplane or a bridge, or too small, as a computer chip. It therefore becomes necessary to reduce or increase the drawing to a size easy to handle and interpret. Scaling allows you to obtain an accurate proportion between the size of the drawing and the actual size of the object.

Scales are indicated on formal engineering drawings in a variety of ways. For example, if the actual size is twice as large as the drawing size, the scale is indicated as half-size, 1=2, or 1:2. If, on the other hand, the actual size is one-half the drawing size, the scale will be double-size, 2=1, or 2:1.

The selection of an appropriate scale depends on the size of the actual measurement to be reproduced in the drawing. For engineering maps and plats, scales of 1"=100' up to 1"=500', or their counterparts in the SI system of units 1:1000 up to 1:5000, are commonly used. The United States Geological Survey maps are prepared to scales of 1:24,000 up to 1:100,000 (see Fig. D3-1). Structural drawings usually have scales varying from 1/4"=1'-0" to 1"=1'-0", depending on the size of the structure with respect to the drawing sheet (see Fig. D3-2). Machine part drawings, on the other hand, are often done full-scale.

The important thing, therefore, is to establish a scale that can be easily implemented on the sheet of paper available to you. For informal engineering drawings, you may find it easier to use a rectangular grid paper, with 1/8" squares, which can be quickly scaled to the size of the object you are drawing.

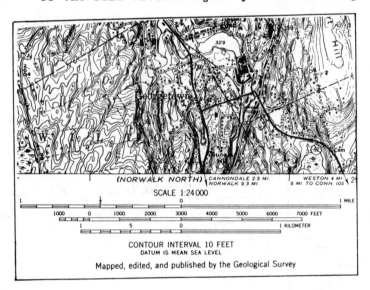

FIGURE D3-1

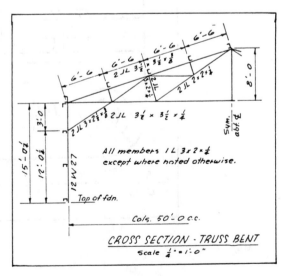

FIGURE D3-2

D3d: DIMENSIONING PRACTICES FOR SCALED DRAWINGS

Dimensions are used in engineering drawings to give an accurate location for points and lines. The dimensions are always given in full size units and are independent of the scale used in the drawing. Figure D3-3 shows a fully dimensioned drawing. Notice the form and location of the dimension lines.

Mechanical drawings have all dimensions necessary for manufacture completely displayed on the drawing. You should never have to measure a distance to obtain a missing size. Structural drawings may occasionally have to be measured, and for that reason the scale used should be prominently shown.

The types of drawings most commonly drawn by engineers are dimensioned sketches used to communicate design ideas (see Fig. D3-4). Design drawings are used to prepare the working drawings needed for the manufacture of a product. With the recent advances in computer-aided design, these drawings can be prepared directly on a computer terminal with graphical capabilities. Section D5 in this manual will elaborate further on this topic.

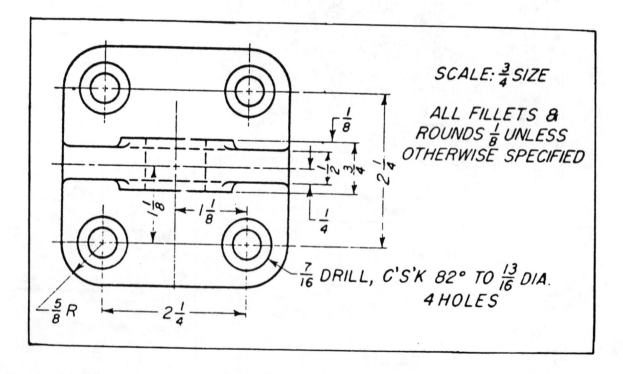

FIGURE D3-3

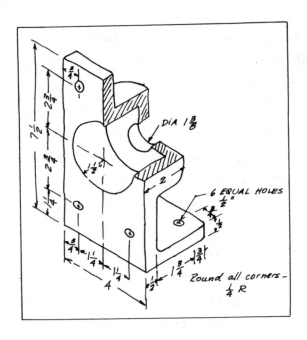

FIGURE D3-4

It is important that you always keep in mind the degree of accuracy needed for a dimension. If, for example, you use a state map to estimate the distance between two cities, your estimate will only be accurate to a fraction of a mile. Maps do not give actual dimensions, and you will have to make use of the indicated scale to reach your value. On the other hand, certain machine parts, such as an engine's piston, may require dimensions accurate to within thousandths of an inch. The accuracy to which a dimension must be defined will depend on the relative size and function of the actual part being dimensioned. Keep this statement in mind when preparing a scaled drawing. If, for example, you are drawing the layout of a room to determine patterns of traffic flow, it would be wrong to indicate the dimensions any closer than the nearest inch.

D3e: INTERPRETING WORKING DRAWINGS

In the process of manufacturing a machine or building a structure, a set of formal scaled drawings is necessary to provide the details needed for production. These drawings are called "working drawings," and consist of two basic types: detail and assembly.

Detail drawings supply all the information necessary for the manufacture of each part, including the name, shape description, dimensional size, and notes detailing material, special machining, finish, or heat-treatment. Figure D3-5 shows an example of a detail drawing. Notice that the shape of the part is not described by means of a pictorial drawing. Shape description in detail drawings is done by means of multiview orthographic projections. The easiest way to understand this type of projections is to imagine that the part has been placed inside a glass box (see Fig. D3-6). The walls of the box become the projection planes. Imaginary projection lines are used to bring the separate views to each projection plane. If you further visualize the unfolding of the glass box, the views appear in what is called an orthographic projection.

Assembly drawings show the working relationship of the different parts of a machine or structure. They are often shown as a pictorial, with each part in the assembly numbered and listed in a separate table. Assembly drawings provide a visual relationship of one part to another and give a list of the parts to be used for a bill of materials. Figure D3-7 shows an example of an assembly drawing.

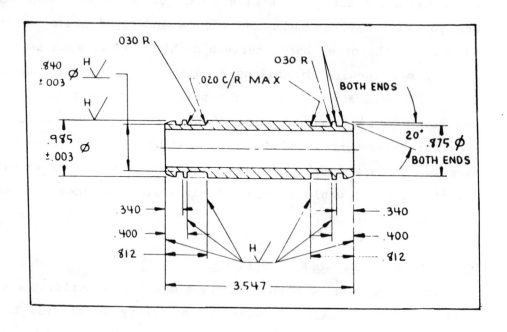

FIGURE D3-5

138

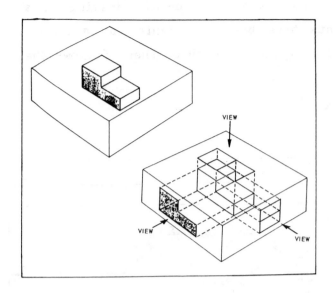

FIGURE D3-6

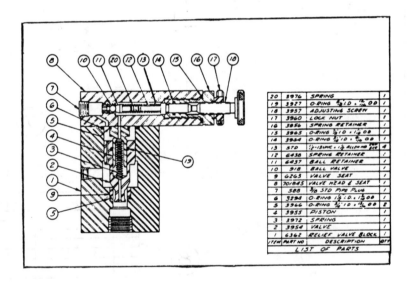

20	3976	SPRING	1
19	3927	O-RING 3/8 I.D. 9/16 OD	1
18	3957	ADJUSTING SCREW	1
17	3960	LOCK NUT	1
16	3956	SPRING RETAINER	1
15	3965	O-RING 7/8 ID 1 1/8 OD	1
14	3964	O-RING 3/4 ID 15/16 OD	1
13	STD	1/2-13UNC x 1 1/2 ALLEN HD SCR	4
12	6438	SPRING RETAINER	1
11	6437	BALL RETAINER	1
10	918	BALL VALVE	1
9	6263	VALVE SEAT	1
8	701845	VALVE HEAD & SEAT	1
7	588	3/8 STD PIPE PLUG	1
6	3294	O-RING 1 1/8 I.D. 1 5/16 OD	1
5	3966	O-RING 3/8 ID 9/16 OD	2
4	3955	PISTON	1
3	3972	SPRING	1
2	3954	VALVE	1
1	6362	RELIEF VALVE BLOCK	1
ITEM	PART NO	DESCRIPTION	QTY
		LIST OF PARTS	

FIGURE D3-7

To be able to interpret working drawings, you must become familiar with standard graphic conventions for shape and size description. Keep in mind that working drawings may show considerable variation in format, but some distinct and required similarities exist among all of them. The more drawings you see, the better prepared you will be to interpret them correctly. Figure D3-8 shows the detail drawing of a shaft on which some major points have been highlighted by a number contained in a circle. A full description of each of these follows the drawing.

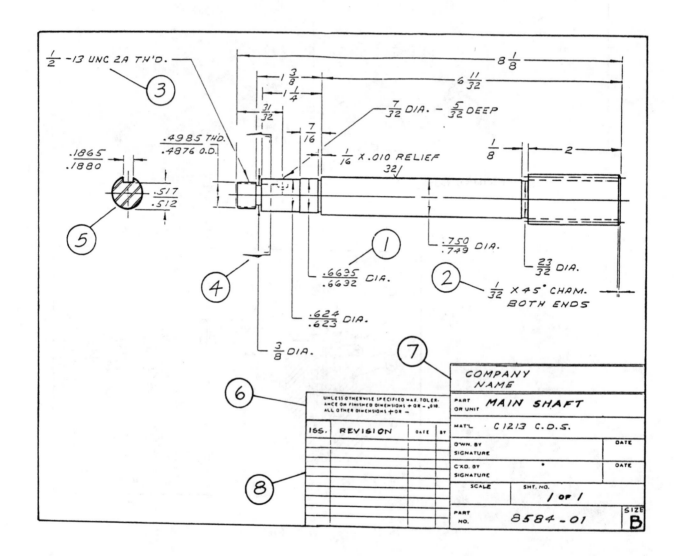

FIGURE D3-8

1 - Limits of size. Extreme permissible dimensions in the manufacture of a part. The variation between these limits is known as the tolerance.

2 - Note indicating the change in diameter at the end of the shaft, known as a chamfer. The fraction represents the width in the horizontal direction and the 45° is the angle of inclination.

3 - Standard note specifying a threaded part. The notations indicate the following thread specifications: 1/2 - major diameter, 13 - number of threads per inch, UNC - thread series symbol, 2A - thread class number and symbol.

4 - Cutting plane line showing the position from which the section view 5 was taken.

5 - Sectional view indicating the cross section of the shaft at the cutting plane. This view was necessary to show the notch on the top of the section. Cross-hatching lines are used to indicate where solid material is being sectioned, and the line design tells that the material is steel.

6 - Tolerance note, indicating general permissible variations in the size of the part. Limit dimensions explicitly given supersede this general note.

7 - Title block, giving supplementary information on the part or assembly. The appearance of the title block varies among companies, but its main features are standard and include: name of manufacturer; name of the part; specific material used; signatures of the draftsman and the checker; scale; sheet number; drawing number, which serves as a filing number and may give information in code form; and sheet size, which is coded size B in the example.

8 - Revision block, which contains a record of the changes that have been made to the original drawing. It includes an issue symbol, which can also be referenced to the field of the drawing next to the change effected; a short description of the change; and date and initials of the person responsible for it.

D3f: REMEMBER

Scaled drawings should be used whenever you have to give a full description of not only the shape, but also the size of an object. Always use an appropriate scale and dimension your drawings to a logical degree of accuracy. Remember that, although working drawings may vary in style, they always follow standard graphical conventions.

CHAPTER D4
TABLES, CHARTS, AND GRAPHS

D4a: ANALYZE FOR BETTER COMMUNICATIONS

Throughout your career you will face situations where numerical information in raw form must be organized for better understanding. In deciding what type of format to use for presentation of these data, be sure to first establish the purposes for which the values were gathered. Once this is done, follow the analysis procedures given in section A to ensure appropriate communication with your audience.

D4b: THE USE OF TABLES, CHARTS, AND GRAPHS IN TECHNICAL COMMUNICATION

The need to present numerical information visually for quick understanding has given birth to standardized practices that are largely used today in all communication situations. Whether preparing a written report or an oral presentation, you may soon find yourself in the position of having to manipulate numbers to facilitate communication with your audience.

Data can be organized in a variety of ways which will be examined in detail in the following sections. Tables, charts, and graphs are important tools for efficient technical communication. They give you a visual model of a set of numbers obtained through observation or experiments. At first, this will help you communicate with yourself by displaying results in a form that enables easy interpretation. Later, you may reproduce them in written or oral reports, so that others have a better understanding of your project. Whatever format you use to display your data, it must be accurate, easy to interpret, and positive in its impact on your audience.

D4c: TABLES

Perhaps the simplest way of presenting numerical information is to use a table of numbers. Different types of information are placed in separate columns, with information related to a single observation placed in rows. Figure D4-1 gives an example of this type of display. Tables are particularly suitable for showing data that does not require

TWIST DRILL SIZES

Number Size Drills

Size	Drill Diameter	
	Inches	mm
1	0.2280	5.7912
2	0.2210	5.6134
3	0.2130	5.4102
4	0.2090	5.3086
5	0.2055	5.2197
6	0.2040	5.1816
7	0.2010	5.1054
8	0.1990	5.0800
9	0.1960	4.9784
10	0.1935	4.9149
11	0.1910	4.8514
12	0.1890	4.8006
13	0.1850	4.6990
14	0.1820	4.6228
15	0.1800	4.5720

FIGURE D4-1

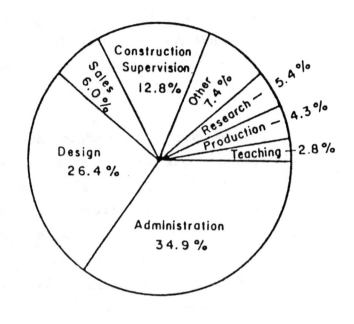

FIGURE D4-2

interpolation, such as the example given in figure D4-1. They are also very useful for cases where plotting would compromise the precision of the numerical values. Tables, however, may be difficult to use when establishing trends or while comparing among values.

D4d: CHARTS

Charts present comparative relationships of numerical data. They may take several forms, depending on the type of data being displayed and the use of the chart. Most charts are designed to show parts of a whole, comparison among different items, or general trends.

To indicate how separate parts relate to the whole, a pie-chart is the most appropriate. It is usually represented as a circle with sectors indicating the separate parts. Figure D4-2 shows an example of a pie-chart.

To show trends and comparison among different items, bar charts are the best suited. Examine figure D4-3 and notice how dramatically it shows the decrease in natural gas reserves during the last decade. Bar charts are very effective for this type of presentation. Two- or three-column bar charts are used to represent two or more individual quantities as they compare to one another and to others (see Fig. D4-4).

Diagram charts are used to show items in a structure or show the relationship among parts in a process. Figure D4-5 is a diagram chart giving a visual description of a complicated process.

Several computer software packages exist today for the efficient production of all types of charts. Whether using these packages or preparing the charts manually, you should observe the following rules:

- <u>Keep the chart simple</u> by limiting the number of elements displayed. Show trends instead of detailed data.

- <u>Balance the display</u> - a formal balance of the displayed elements is normally used for charts. (Refer to section D1k in this manual.)

- <u>Use texture with care</u> - avoid the use of very bold patterns that may distort the manner in which information is perceived. Values and directions of patterns should be similar.

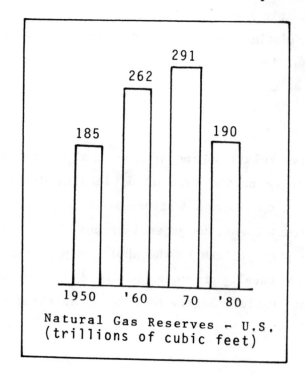

Natural Gas Reserves - U.S.
(trillions of cubic feet)

FIGURE D4-3

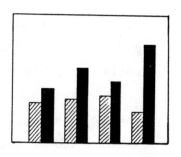

FIGURE D4-4

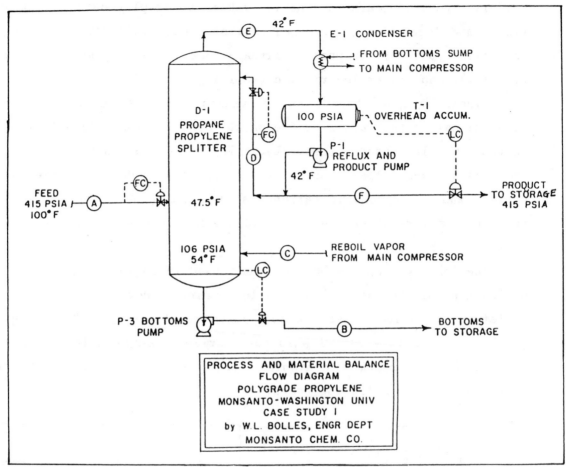

E-1 CONDENSER

42° F

FROM BOTTOMS SUMP
TO MAIN COMPRESSOR

100 PSIA

T-1
OVERHEAD ACCUM.

D-1
PROPANE
PROPYLENE
SPLITTER

P-1
REFLUX AND
PRODUCT PUMP

42° F

PRODUCT
TO STORAGE
415 PSIA

FEED
415 PSIA
100° F

47.5° F

106 PSIA
54° F

REBOIL VAPOR
FROM MAIN COMPRESSOR

P-3 BOTTOMS
PUMP

BOTTOMS
TO STORAGE

PROCESS AND MATERIAL BALANCE
FLOW DIAGRAM
POLYGRADE PROPYLENE
MONSANTO-WASHINGTON UNIV
CASE STUDY I
by W.L. BOLLES, ENGR DEPT
MONSANTO CHEM. CO.

Courtesy Professor Buford Smith, Washington University.

FIGURE D4-5

- <u>Avoid grid lines</u> on bar charts if used to give comparisons and establish trends.

- <u>Place titles and labels carefully</u> - refer to section D1j in this manual.

D4e: GRAPHS

Most engineering data is presented in the form of line graphs. These graphs are more exact than the charts described in section D4d. They illustrate trends, help predict values, and are extremely useful in the accurate presentation of results of test data obtained in experiments.

145

Two types of scales are normally used to display data in a graph: linear and logarithmic. Combinations of these scale types give three possible types of graphs: linear-linear, logarithmic-logarithmic (log-log), and linear-logarithmic (semi-log).

Linear graphs have the distances on the coordinate axes proportional to the values of the data. Usually, the origin is located in the lower left-hand corner. Variables for the ordinate (y-axis) and the abcissa (x-axis) are chosen showing the independent variable as the abcissa and the dependent variable as the ordinate. This relates to the functional relation $y = f(x)$. Figure D4-6 gives an example of this type of graph.

The log-log graph is very commonly used in engineering. This type of graph should be used to express logarithmic functions or data values satisfying equations that contain power relations ($y=ax^n$). Log-log graphs are also very useful when both variables cover a large range of numbers and it becomes convenient to compress both sets of data. Figure D4-7 shows a log-log graph.

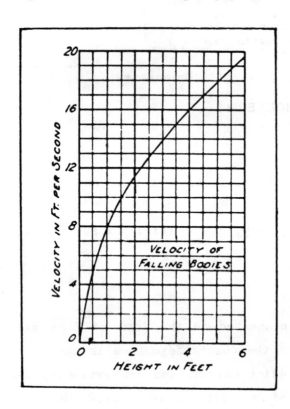

FIGURE D4-6

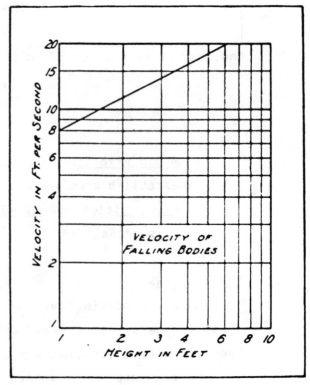

FIGURE D4-7

Semi-log graphs, where only one of the variables is represented in a logarithmic scale, are less often used in engineering. They are useful when the data behaves in an exponential form ($y=be^x$) or when only one of the variables covers a large range of numbers. Figure D4-8 is an example of a semi-log graph.

The choice of the correct scales is very important in any graph so that a reasonable impression is given to the viewer on the degree of change in the variables. For ease in interpolation, you should make the value of the smallest division in the scales a power of 10 times 1, 2, or 5. In other words, don't put 4 or 6 small divisions between numbered lines 20 and 30 for example, unless there is a very specific reason for doing so.

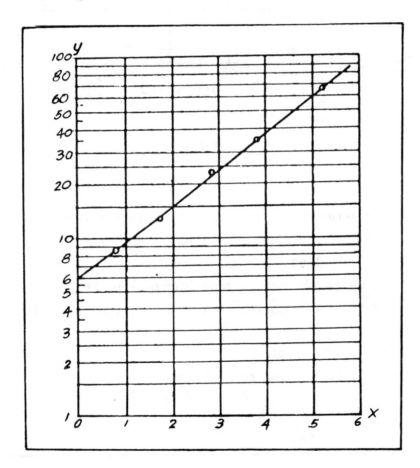

Graph for equation $y = be^x$

FIGURE D4-8

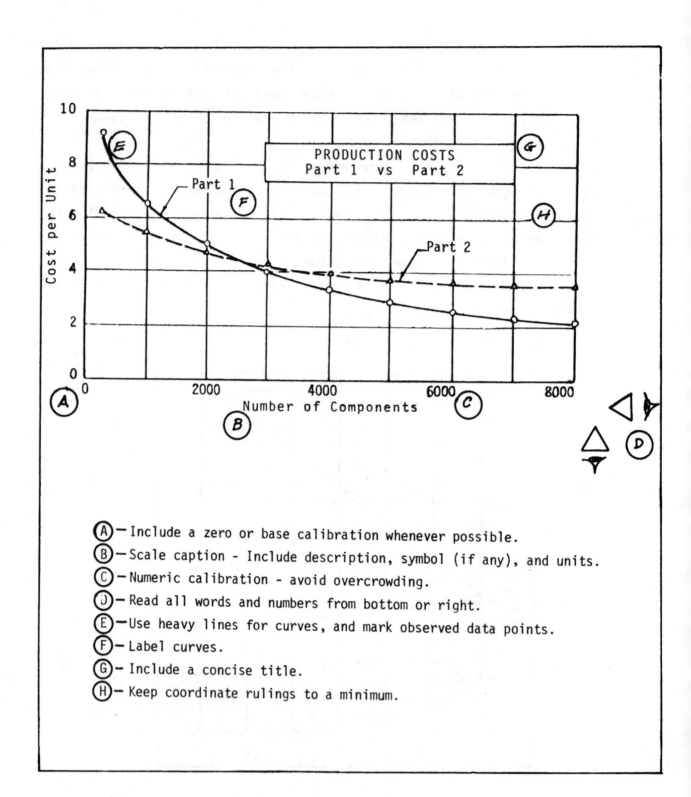

PRODUCTION COSTS
Part 1   vs   Part 2

Ⓐ — Include a zero or base calibration whenever possible.
Ⓑ — Scale caption - Include description, symbol (if any), and units.
Ⓒ — Numeric calibration - avoid overcrowding.
Ⓓ — Read all words and numbers from bottom or right.
Ⓔ — Use heavy lines for curves, and mark observed data points.
Ⓕ — Label curves.
Ⓖ — Include a concise title.
Ⓗ — Keep coordinate rulings to a minimum.

FIGURE D4-9

Figure D4-9 gives a graphic summary of the standards commonly followed for graph plotting. Examine this figure carefully to become acquainted with the most important rules to follow in graph making. Commercial graph paper is available in the most common grid patterns, which include log-log and semi-log. Rectangular grids vary from 4,5,8, or 10 lines per inch, and 5 and 10 lines per centimeter.

Today much graphing is done by the use of appropriate computer software. Depending on the type of software available, you may or may not have much control over the graph display. Section D5 in this manual will elaborate further on this topic.

D4f: <u>REMEMBER</u>

Whenever presenting numerical results, use a visual model to facilitate communication. The format chosen to display the data may vary, but should always be accurate and easy to interpret.

CHAPTER D5
COMPUTER GRAPHICS

D5a: ANALYZE FOR BETTER COMMUNICATIONS

The advent of computer graphics has revolutionized the way in which graphical communication occurs. It has not, however, changed the fact that you, as an engineer, will still have to answer all the basic questions relating to this type of communication. The computer will simply act as an extra and powerful tool to help you make your decisions. Careful study of your objectives and an analysis of your audience's needs becomes more important than ever when making use of computer graphics methods.

D5b: BASIC CAPABILITIES OF COMPUTER GRAPHICS SYSTEMS

Computer graphics can be thought of as a collection of computer hardware and application programs directed to one specific goal: having pictures drawn by a computer. Engineers and scientists have always used drawings to communicate their ideas, but the advent of computer graphics has given them a simple and precise method for the creation and manipulation of these drawings.

In general, computer graphics systems have the following basic capabilities:

- Allow fast, easy input of data.
- Store large volumes of data.
- Allow quick display of data in graphical form.
- Permit manipulation of the graphical display.
- Have excellent performance on repetitive tasks.
- Allow quick production of output.

D5c: COMPUTER GRAPHICS HARDWARE

Computer graphics can be performed in both large scale and personal computer systems. Major hardware components are shown schematically in Figure D5-1. These components can be generally divided into the following categories: input devices, computers, and display and output devices.

The computer is the main component and can be a mainframe or a microcomputer, depending on the speed of the processor and addressable memory capacity.

Input devices are used to communicate with the computer and control the creation of a picture. The most common type of input device is the keyboard, which includes standard typewriter keys and several special function keys for specific actions, such as controlling the screen cursor (see Fig. D5-2). Other input devices commonly found in graphics systems allow the user to communicate to the computer the location of a particular point. Among these are digitizers, joy-sticks, light-pens, and others. Figure D5-3 shows some of these devices.

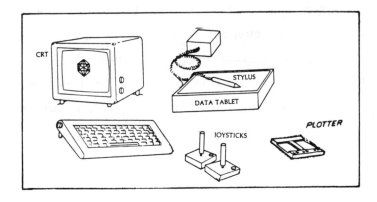

FIGURE D5-1

FIGURE D5-2

FIGURE D5-3

Most modern graphics systems use a cathode-ray tube (CRT) with the capability of displaying pictures as an output device (see Fig. D5-4). These graphics terminals vary depending on the mechanism used to display the image, the resolution on the screen, and their ability to display color. The plotter is another type of output device. Depending on the scheme used to form the picture, plotters can be classified as pen, electrostatic, thermal, and printer plotters. Figure D5-5 shows some of these.

D5d: GENERAL APPLICATION PROGRAMS

Applications of computer graphics occur in all areas of engineering. The following are some major applications in the technical communications area:

1. Computer-generated text. Computer text for drawing or presentation output differs from the standard alphanumeric input typed on your computer keyboard. Usually the lettering types are generated by software included in a computer-aided drafting package. However, if you are interested in producing

FIGURE D5-4

FIGURE D5-5

signs, transparencies, or slides, various types of software are available in the market for the sole purpose of generating different letter-styles. In most software packages the lines of text are entered in the computer terminal directly from the keyboard, just as you would type them in a typewriter. You then command the system to display the text in a chosen type-style at a particular location on the screen. The software automatically establishes the appropriate spacing for letters and words. Figure D5-6 is an example of computer-generated text.

2. <u>Computer-aided design (CAD).</u> Computer graphics is very useful when you look for alternative solutions during the design stage of a product. If production of a 3-D machine part image from various angles is required, this can be rapidly accomplished by means of a 3-D computer modeling package. Automotive and aircraft engineers use the techniques of CAD to establish the contours of various surfaces commonly encountered in their industries. Electrical and electronic circuits can be easily designed by CAD methods. Symbols representing the different components are added one at a time until the correct configuration is obtained. Figure D5-7 shows output obtained from computer-aided design packages.

Amount of Water Delivered to a Stream
For Three Different Rainstorms

AMOUNT OF WATER DELIVERED TO A STREAM
FOR THREE DIFFERENT RAINSTORMS

FIGURE D5-6

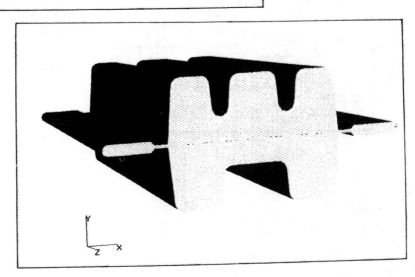

FIGURE D5-7

3. <u>Computer-aided drafting</u>. These types of systems are available today for the fast and efficient production of drawings. Most drafting packages are menu-based for easy manipulation and permit automatic choice of units and scales. Once the drawing is created and stored electronically, modifications to the original can be made very quickly. Figure D5-8 shows examples of drawing output from computer-aided drafting systems.

4. <u>Computer-generated charts and graphs</u>. Data plotting can be easily accomplished by computer graphics methods, and a wide range of software is available for this purpose. As a user, you can choose the type of chart or graph needed to display your data. Data values are entered in a format established by the software, as are all labels and titles. Usually, most other tasks are performed automatically by the system. Figure D5-9 is an example of computer-generated charts and graphs.

D5e: <u>REMEMBER</u>

Do not rely on the computer graphics system to answer basic questions on your design or technical drawing. Think of it an an additional and very powerful tool which can simplify your task when communicating graphically.

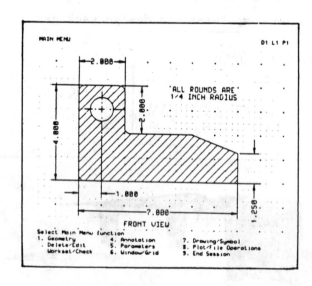

FIGURE D5-8

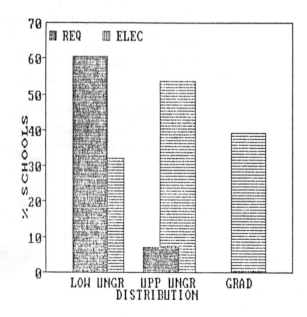

FIGURE D5-9

154

CHAPTER D6
PHOTOGRAPHY

D6a: ANALYZE FOR BETTER COMMUNICATIONS

As an engineer, you must communicate with others effectively and efficiently. When preparing written or oral communications you may find that words, charts and graphs, sketches, and even elaborate drawings do not convey your message accurately or effectively. A picture, or a series of pictures, to supplement written or oral communication can make your communications easier to prepare and understand.

You have been urged in Section A to begin and end every communication with an analysis of the participants and the situation. This analysis should give you direction as to how photography may improve the communication process. While it has often been said that "A picture is worth a thousand words," you must recognize that the decision-making process is necessary if you are to take full advantage of this communication medium.

D6b: ESTABLISH OBJECTIVES

In using photographs, the engineer must keep in mind the function for which they are being used. Photographs are used purposefully:

1. to show steps in processes or time sequences,
2. to verify technical details,
3. to introduce broad topics of general interest,
4. to elaborate on problems under study,
5. to motivate and promote interest.

Never just use a photograph and hope it will be effective. If the photograph doesn't serve a purpose, then do not include it.

D6c: ANALYZE THE EQUIPMENT AND SUPPLIES

A decision to use a picture forces you to consider what equipment and supplies are needed to secure the photographs. The decision process must focus on the type of camera, variety of lenses available, the film, and the sources of light available for taking the photographs.

<u>D6d</u>: <u>Cameras</u>

Cameras, regardless of type, are essentially the same and provide a way of holding the film, controlling the light, and focusing the image on the film (see Fig. D6-1). Three cameras with which engineers should become familiar are the simple box or Instamatic camera, the instant picture or Polaroid camera, and the 35mm reflex camera.

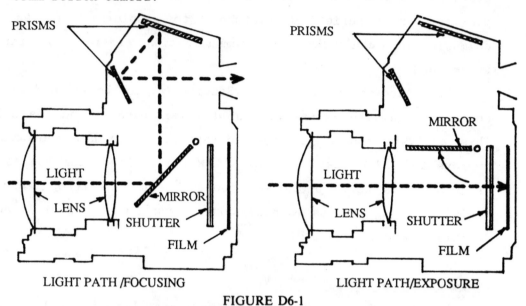

PRISMS

PRISMS

MIRROR

LIGHT

LENS

MIRROR

SHUTTER

FILM

LIGHT PATH /FOCUSING

MIRROR

LIGHT

LENS

SHUTTER

FILM

LIGHT PATH/EXPOSURE

FIGURE D6-1

For the person who does not have a good knowledge of photography, the <u>simple instant load camera</u> may be the best choice. The Instamatic camera uses a film that is preloaded in a cartridge, and when properly inserted, the camera is ready for the first exposure. The camera will automatically control the exposure to the general light conditions, and you will only have to focus the image so that it appears sharp in the viewfinder. In most models the camera will warn you when the flash unit is needed to provide necessary light.

The <u>instant picture camera</u>, first introduced as the Polaroid Land Camera, is designed to make the picture taking process easy and to give you a recorded image print within seconds of your exposing the film. This camera's capacity to give hard copy, while still on site, allows you to evaluate each print and, if need be, reshoot the scene.

The 35mm Reflex Camera, particularly the 35mm Single Lens Reflex Camera, is the preferred one for quality general photography. It allows you to control the exposure and focus to create the picture that best serves your needs and when combined with the proper film, lens, and filters can give photographs of exceptional quality.

D6e: The Lens and Light

When making photographs you will be using the camera lens to collect and direct the light to the film. If you select the simple instamatic camera, your lens probably will be a "fixed;" however, if you select a camera that does provide variety of optional lenses and lens characteristics, the following discussion will be helpful.

Focal Length: The focal length of a lens is the distance from the lens to where the image of a distant object is formed. A normal lens allows you to produce a picture approximately as it would look to the human eye, and the focal length would be roughly equal to the diagonal of the film cell being used.

Wide Angle Lens: The short focal length lens, often called the wide angle lens, allows you to incorporate more of the scene into your picture. You can use the wide angle lens relatively close to your subject and still not lose the surrounding elements, but it does tend to exaggerate the distance relationship between objects which are in front, or behind, the focal plane.

Telephoto Lens: The long focal length lens, often called telephoto lens, allows your camera to function as a telescope. The picture you see through the viewfinder is the magnified image that will be recorded on the film. However, the long focal length lens gives the picture a flat quality, and the elements in the picture seem closer together than they really are. It also makes objects out of focus if they are too close to the camera or too far behind the focused object. This can work to your advantage by deliberately fuzzing out unwanted background elements.

F Numbers: The f/number is a numerical way of expressing the relationship that exists between the focal length of the lens and the actual lens opening (see Fig. D6-2). The lens opening may be regulated by a diaphragm placed between two lens components. The f/number is a simple fraction of the len's focal length. A lens set at f/2 has a lens opening diameter which measures 1/2 of the len's focal length.

FIGURE D6-2

Courtesy Minolta Corporation.
Remember that the smaller the f/number, the larger the opening in the lens, and therefore the faster the lens (more light is allowed to strike the film per unit of time). When you move the diaphragm by one f/number, you double the amount of light available through the lens. When you increase the f/stop by one f/number, you reduce the amount of light by 1/2. (See Fig. D6-3)

f-	1.4	2.8	4	5.6	8	11	16	22	32	45	64
Fraction of Focal Length	$\frac{1}{1.4}$	$\frac{1}{2.8}$	$\frac{1}{4}$	$\frac{1}{5.6}$	$\frac{1}{8}$	$\frac{1}{11}$	$\frac{1}{16}$	$\frac{1}{22}$	$\frac{1}{32}$	$\frac{1}{45}$	$\frac{1}{64}$
Exposure Ratio	$\frac{1}{4}$	$\frac{1}{2}$	1	2	4	8	16	32	64	128	256

FIGURE D6-3

D6f: Film

You have a variety of photographic film from which to choose. However, most decisions will focus on speed, graininess, and color sensitivity.

Speed: Film speed is a description of how sensitive the film is to light. A fast film (ASA 400) means that a picture can be taken with less light and allows you greater range of shutter speed and aperture settings. A slow film (ASA 32) requires more light to take a good picture and therefore limits your options on shutter speed and aperture settings.

Graininess: When you select film speeds you also influence the graininess of your photographs. As the film speed increases the graininess increases, and this is most apparent in pictures that have been greatly enlarged. On smaller prints graininess does not present major problems. When you want photographs showing very fine detail, or you plan to enlarge the prints to 8" x 10", use a relatively slow or medium speed film.

Color Sensitivity: In general, film used for black and white photography may be classified as blue sensitive, orthochromatic, or panchromatic. It is important to understand that the film sees only the relative brightness of the scene, while the eye responds to both brightness and color.

Panchromatic film is most often used in general photography and is readily available at your local supplier. The film is sensitive to all colors and when properly filtered, reproduces the blacks and whites in a manner that looks quite natural.

D6g: ANALYZE THE EXPOSURE

How well you do the exposure, which really poses two distinct problems, will determine the quality of your picture. The first problem is setting the shutter speed and the aperture opening; the second is the type of image wanted.

D6h: The Shutter

The camera shutter mechanism controls the time relationship in the exposure. The shutter speed-scale settings usually offer a choice of timed exposures ranging from 1/500 of a second to 1 full second in steps which double the time. The markings on the shutter speed-scale may be 1, 2, 4, 8, 15, 30, 60, 125, 250, 500. The number 1 stands for 1 full second; all others are fractions of a second: 2 is ½ and 4 is ¼, and so forth (see Fig. D6-2).

The scale also may include T and B markings. The T stands for a time exposure, and when the shutter speed-scale is set at this marking, the shutter will stay open until you trip it again. When the shutter is set on B, the shutter will remain open as long as your hold the shutter button down.

D6i: The Aperture

With time controlled by the shutter, you can control light intensity by selection of the aperture setting. As the size of the aperture increases, the intensity of the light increases. The size of the aperture is designated by the f/number. Remember that the smaller the number following the f/stop, the larger the opening. (See D6e)

The normal f/number series is like the shutter-speed scale in that each number gives one half the intensity of the preceding number and twice the intensity of the next number. Since the area of a circle (the aperture) is proportional to the square of the diameter, doubling the diameter quadruples the area; therefore, increasing or decreasing the diameter by a factor of 2 increases or decreases the area by the factor of 4 (see Fig. D6-4).

FIGURE D6-4

160

D6j: Shutter and Aperture Related

The f/number scale and the shutter speed-scale are related in that each number change gives you one half or double the light intensity or time of the previous setting. Because of this arrangement, changing the shutter speed in one direction and changing the aperture in the opposite direction will mean no change in exposure intensity. However, the image that you record will not be the same because of changes in depth of field.

Should you use f/22 at 1/60 or f/8 at 1/500? Decide on the basis of what sort of image you want in the final picture. The aperture of f/22 will give you relatively great depth of field, while f/8 will give you a shallow depth of field. The shutter speed set at 1/60 will not stop a fast-moving object, but the shutter speed of 500 will give sharp images of most moving subjects. Make the decision based on the image you want.

D6k: Exposure Guides

The tables and charts supplied by the film maker indicate shutter speed, the light conditions of the day, and the characteristics of the scene being photographed. These are simple to use and quite reliable under the conditions given on the material.

Today the simplest and most accurate aid is the photoelectric meter. Many types of meters are available, but the most convenient ones are those built into the camera. When using a camera equipped with a built-in light meter, be sure to familiarize yourself with the general operating procedures before loading the film.

D6l: REMEMBER

The user of the camera is much more important than the camera itself. You are the most important part of any photographic process. Your eyes, your mind, your understanding, and your skill make your photographs unique. Build on your knowledge of the camera, the physics of light, and the chemistry of film. Then your ability to communicate is increased, your effectiveness sharpened.

CHAPTER D7

PHOTOCOPYING

D7a: ANALYZE FOR BETTER COMMUNICATIONS

When you prepare written communication material, recognize that part of the material may need to be reproduced. Written laboratory reports, proposals, technical papers, or handout materials in oral programs all require that you consider if and how the materials will be reproduced. Become familiar with the operation and availability of the copier processes in your organization so that your reproduced materials present your work in the best possible way.

D7b: ESTABLISH OBJECTIVES

At this time review the objectives for your communication material and answer the following questions: How many copies will be required? Will the material to be reproduced include illustrations, photographs, or charts? Does the written material require that a paste-up (pieces assembled into a single sheet) be used as the original for reproduction? Does the communication material include multiple colors?

D7c: ANALYZE FOR EQUIPMENT TO BE USED

Several methods are used to reproduce multiple copies of your typed, hand-lettered, hand-drawn, printed, or photographed presentation items. The most common are spirit duplicating, mimeographing, photocopying, and printing. In modern offices the two processes most used are photocopying and printing, due to their speed, quality, and cost.

D7d: Photocopying

Recent developments in photocopy processes and equipment have done much to aid in the presentation of valuable verbal and visual material. The modern office would be much different without the photocopy machine.

Two related processes below are often referred to as photocopying processes. Both require charged surfaces that are affected by light. The charged surface will then hold a substance (toner) to create the image.

In xerography, a plate or drum that is photoelectrically sensitive is given an electrical charge. The original copy to be reproduced is exposed to light with the white area reflecting light to the charged surface. The light that reaches the charged surface dissipates the electric charge, leaving a charge only in the image area. The copy paper (ordinary bond paper) is charged, and a toner (fine colored powder) is transferred to the paper from the charged plate or drum. The paper is then heated to fix the powder permanently to the image area of the paper.

A related process, the electrofax process (RCA trademark name), requires a special copy paper coated with zinc oxide. The copy paper is given a uniform electric charge; then the image of the original is placed on the charged copy paper by reflective light to dissipate the charge. Where no light is reflected, the charge remains and retains the image when dusted with toner. The toner is then fused to the paper with heat.

The photocopy machines are easy to operate but are an expensive reproduction process if a large number of copies are required. With care in preparation of the original, the copies produced are of good quality.

D7e: <u>Printing</u>

If your objectives include a need for more than 100 copies, or multi-color reproduction, consider using the offset printing process. The process provides excellent reproduction quality at a comparatively low cost per copy and allows multiple color reproduction of excellent quality.

If you select printing as the method of reproduction of your copies, then get the printer's advice on the preparation of the copy before doing a final version. The printer can give you press limits of text area, paper size, hints to minimize cost of printing, and specifics that the job may demand.

<u>D7f</u>: <u>PHOTOCOPY PROCEDURES</u>

Photocopying a page or multiple pages of a typed report is a relatively simple task. Modern copying machines are very operator-friendly with instructions provided to follow. The following procedures are presented to allow you to reproduce your written communications that include not only text, but also illustrations, photographs, charts, and assembled text. Don't forget that many copying machines are able to enlarge or reduce selected areas.

<u>D7g</u>: <u>Copy</u> <u>Preparation</u>

When preparing text copy, make sure you have a clean typewriter and a good ribbon or that your printer is giving good density for each character.

Use white paper for making your originals. Remove all paper clips and staples from copy before you begin the photcopy operation.

Be sure the copy is as free as possible of smudges and marks.

<u>D7h</u>: <u>Developing</u> <u>an</u> <u>Attractive</u> <u>Layout</u>

If your material to be copied includes a combination of elements (type-illustrations-charts), sketch your arrangement of the elements on the page.

Maintain a relationship of elements that allows for ease of reading of related parts.

The arrangement of elements should maintain page shape balance and provide a visual image that seems to fit all the parts into a pleasing arrangement.

Use sharp scissors when cutting the elements that you will assemble and hold the scissors so that the cut edge forms an incline toward the base sheet.

When putting the element onto the page for the paste-up, use a glue stick or rubber cement to position each element in place.

<u>D7i</u>: <u>Line</u> <u>Artwork</u>

When your design includes line art work, it is best to make the drawings with a reproductive black pen. Drawings done with pencil often tend to smear and usually do not have the even density necessary to copy as a solid line. If you do wish to use penciled drawings or sketches, copy the original pencil work and then use the copy to make others.

D7j: Photographs

The photocopier does not give a good reproduction of unscreened continuous tone photographs. But screened photographs, such as those appearing in newspapers or magazines, will reproduce quite well. If you are not able to screen (halftone) your pictures, most local quick-copy services can provide you with suitable positive halftones of your photographs for a nominal charge.

D7k: Special Techniques

In preparing your originals for photocopying, keep in mind that variation in type and line values represented on the layout can increase the impact of your presentation.

D7l: Two-Sided Copying

Most photocopying machines will only copy one side of your original. If you wish to save paper used, it is possible to copy on both sides of your copy paper. The process requires that after the first side of the page has been copied, the paper be removed from the off-feed table and re-inserted into the magazine for new paper. When inserting the paper in the magazine, position the paper plain side up, and the copied side top down or reversed of the original being copied. Run one copy to make certain that the orientation of second side copied is correct.

D7m: Overhead Transparencies

When you determine a need for an overhead transparency, you can use the photocopying machine for transferring the image from the paper original to the plastic overhead transparency material. When using the photocopier for making overhead transparencies, you will need a special coated electrically-charged and light-sensitive film. Be sure to match the film type to the machine. The process for using the photocopying machine is the same as copying on paper. Before loading the plastic sheets into the magazine, fan the edges of the plastic sheets so that only one sheet will be fed through the machine per cycle.

Thermal-type machines are also available for transparencies and a different type of film is required. Learn how to use the equipment available and make sure that you have the correct film for your process.

D7n: PRINTING

The decision to have your copies produced by offset printing usually removes you from the actual printing (copying) process. The steps related to copy preparation, art work, and paste-up are your major areas of concern.

D7o: Copy Preparation

When the originals to be reproduced include only type, then it is possible for you to type directly onto paper master plates that go directly to the printing press. These direct-image plates are inexpensive and give you a means of securing up to 1000+ copies at a most economical cost.

Electrostatic Plates allow you to use the photocopying process to transfer the image (type and drawing) to paper or foil printing plate. This process allows for reproduction of 1000+ copies of excellent quality at a relatively low cost per copy.

Metal Plates give the advantage of longer runs (more copies) and higher quality. The photographic process allows for very accurate transfer of image from original to the plate. Therefore, nice quality, fine detail, and picture tones are maintained from copy to copy. Due to the added step involving photographic film and chemicals, this process does add to the cost of each copy.

<u>D7p</u>:   <u>Paste-Up</u>

The paste-up (assembly of art, type, and photographs) for reproduction is a necessary step in offset printing process. The processes discussed for preparing copy for photocopying should be reviewed. Your paste-up should accurately position each element as it will be printed on the reproduced page. If multi-color printing is planned, overlays for each color must be made in position to the paste-up so that the various colors are located in position on the reproduction. If photographs are included, they will not be attached to the paste-up. You will make windows (blocked-out areas) on the paste-up so that the photographic halftone negatives can be positioned prior to making the printing plate.

<u>D7q</u>:   <u>Art</u> <u>Work</u>

The characteristics of art work discussed in section D7i are appropriate for printing reproduction processes. However, due to the photographic reproduction process that can be used in conjunction with the printing process, the art work can be more complex and detailed than you would use for the photocopying processes. Therefore, the art work can employ greater structural detail and incorporate multiple tones of a single color or multiple colors.

<u>D7r</u>:   <u>REMEMBER</u>

As an engineer, you should develop some understanding of how your written communication materials can be photocopied. When you anticipate the need for photocopying, you minimize problems that affect the quality, readability, and cost of reproducing your communications. A understanding of what the photocopier can do and how you put the message together to enhance reproductability can improve the effectiveness of your communications.

CHAPTER D8
VISUAL AIDS

D8a: ANALYZE FOR BETTER COMMUNICATIONS

Visual aids can be your most important communication tool. They assist you in achieving two major communicative purposes - to _inform_ by giving general information or to _instruct_ through specific information. Use of well-planned and well-executed visual aids will greatly enhance your presentation and your effectiveness as a technical communicator.

D8b: ESTABLISH OBJECTIVES

In planning visual aids, limit yourself to a few concisely stated achievable objectives. Build upon an idea, a generalized statement of purpose. Remember that each communication does have a purpose and that any visual aid used should move toward that objective. Don't fall into the trap of using audiovisuals because they're expected or seem appropriate. Develop your communication plan, and then work your plan.

D8c: OVERHEAD TRANSPARENCIES

Overhead transparencies can easily present a variety of concepts, ideas, facts, outlines, and other visual images to a group. Because the room lighting is constant and since you can maintain eye contact with your audience, this device is an effective tool.

D8d: Planning

When considering the use of "overheads", keep in mind the recommended planning sequence for all communications. Consider the objectives of your presentation and identify the purpose of each transparency. Prepare an outline of the content to be included and make sketches to show the arrangement.

D8e: Preparing Transparencies

Many processes have been developed for making the overhead transparency. These range from simple hand lettering or drawing on acetate to purchased printed and framed transparency cells. Self-made transparencies can be created on clear acetate using marking pens, tape, dry transfer letters, or adhesive cutouts, but higher quality can be obtained by reproducing prepared diagrams or copy on either heat-sensitive or electrostatic film.

When you prepare copy for the transparency, keep in mind these guidelines:

a. Limit the content to a single concept.

b. Keep within a 7½" x 9½" area for compatibility with the 10" x 10" projector surface and 8½" x 11" film.

c. Arrange the format of the copy so that the viewer can see all parts of the image.

d. Provide bold lettering by using a display typewriter or by enlarging typed copy on a copy machine.

e. If adding color, use <u>sharp-tipped</u> felt pens to write or draw lines and <u>broad tip</u> felt pens to color areas with minimum stroke overlap.

f. If using overlays, make an original of each overlay and then make a registration mark (+) in each corner of the composite master. This will allow you to align the overlay when mounting for projection.

Hand-Drawn or Lettered Clear Acetate Transparencies

Preparing transparencies on acetate is a simple and easy task. However, the transparencies are usually of marginal quality and not permanent. This method is best used as a replacement for the chalkboard or for quickly supporting oral remarks with ideas, sketches, or process flow.

The materials needed are clear acetate (.005-.010" thick), felt tip markers in various colors, and a cardboard mount if you want to have a more controlled work surface. To improve the quality of the hand-drawn transparency, consider using transfer letters or colored tape. Charts, graphs, and other illustrations can be quickly made using these materials, but use a grid paper or a special overhead projection tablet under the clear film to guide placement of letters and tape. Avoid cutting the elements while in position on the acetate, since the small indention in the acetate will project as a dark line.

Making Transparencies Using Heat-Sensitive Film

Transparencies can be made quickly and inexpensively using heat-sensitive film if you have a infrared-light copy machine. The materials needed are the heat-sensitive film, the infrared-light source, and a printed or drawn paper original. Set the control on the machine to the correct setting, usually marked "transparency." Place the heat-sensitive film on top of the paper original and feed into the machine. When the film and paper feed out of the machine, separate the two. Now you are ready to mount your transparency.

In preparing the original for the heat-sensitive transparency process, use black carbon ink, a new carbon or cloth typewriter ribbon, or a copy of your original from a photocopy machine. Originals in ordinary pen or pencil will usually not reproduce on the film.

Making Transparencies on Electrostatic Film

The modern photocopying machine provides an excellent method of making transparencies. The materials needed are the coated light-sensitive film, the paper original, and the electrostatic photocopier. The process of making the transparency is the same as making a photocopy except you need to insert the film into the paper magazine.

The electrostatic film is available in black and in various colors, including a full-color film for reproducing continuous tone color originals. Follow the film manufacturer's instructions.

D8f: The Overhead Projector

Use one or more of the following projector techniques to make your presentation more effective:

a. Use a _pointer_ to locate items _on the transparency_ rather than pointing to the screen. This keeps the viewers' attention on the screen and not on you.

b. Control the rate of information presented by covering the transparency and then selectively removing the cover to disclose the image under discussion. This progressive disclosure technique presents a logical sequence and provides key points for your oral presentation.

c. Use overlays of transparent sheets to build your image. This allows you to discuss separate elements and also put them together to present the total structure, process, or organization.

d. Color in elements as you make your presentation. Use the colored felt marker to highlight the items you are discussing.

e. Use a three-dimensional item to silhouette a part or an arrangement. This allows you to manipulate parts and show features of various arrangements.

f. Simulate motion by using special polarized plastic adhesive film and a polaroid spinner attachment. This allows you to show current flow through a circuit, fluid movement in systems, or other motion elements.

g. Adapt simple opaque paper cut-outs to discuss plant layouts, systems arrangements, and organizational planning charts.

D8g: PHOTOGRAPHIC SLIDES

Photographic slides can help convey general or specific information or help in developing attitudes and values regarding your topic. The photographic slide gives a real representation of actual field or laboratory conditions, specific design problems, elaborate equipment in use, or other situations difficult to describe in an oral presentation. If you plan the slide presentation well, it can give you a valuable communication tool.

D8h: Analyze the Equipment and Supplies

Before planning your slide programs, review Chapter D6 - Photography. That chapter describes camera usage, which is most important in making good slides.

Complexity of slide preparation must be considered. The following list will help you select the equipment and supplies needed:

a. How will the slides be taken?

b. Will the slides be taken indoors or outside?

c. Will a tripod be needed for the camera?

d. What lens(es) will be used on the camera?

e. Will a flash be used on the camera?

f. Will color or black and white film be used?

g. How will the slides be processed and mounted?

h. Will copy shots be made?

i. Will title or text slides be used?

j. Will the slides be used in conjunction with an audio tape?

k. Will the tapes and slides be synchronized?

l. Will multiple projectors be necessary?

m. Will a remote projector control be available?

D8i: Photographic Equipment and Supplies

The cameras described in paragraph D6d can be used to photograph your slides. A camera with adjustable lens settings, shutter speeds, and focusing is especially useful to provide the flexibility needed under a variety of light and action conditions. You may also need a hand-held light meter, tripod, flash unit or photoflood lights, and a cable release to satisfy all requirements.

Color film giving a positive image is generally used for slides. However, it is possible to make slides from negative film; the processed negative film is used to make positive images by exposing other film. A special Polaroid film is also available for instant slides.

A major cost in preparing a slide program is film developing. If you plan more than three rolls (36 cells each) of slide film, developing kits and slide mounts are available at most camera shops.

If you plan to copy material or take time sequence shots, then a tripod or a copy stand should be available. These are particularly useful in preparing title and text slides.

D8j: <u>Planning</u> <u>Your</u> <u>Presentation</u>

Your preliminary planning has identified your audience. Now give particular attention to the content of your slide series.

As you develop visuals, try to "see" the content you wish to present. Remember, you are preparing a visual aid for an oral presentation. When you write or speak you are concerned about words, but in preparing visuals you need to think of pictures - <u>visuals</u>. You should develop specific ideas on what to include in each part of this visual presentation.

The following sequence of steps will help you develop plans for presentation slides:

- Prepare an outline of the contents.
- Create a sketch or series of sketches that relate to the content identified.
- Make up a card (3" x 5" or 5" x 7") for each sketch and leave enough room for narration notes and specific photographic directions.
- Arrange the cards to follow the outline. You may want someone to review and comment on the arrangement and content of sketches.
- The arranged sketches along with your narration notes become your "storyboard" for the proposed slide series.

Keep in mind the length of your presentation. The contents of any visual definitely affect the time required to present the material. You should estimate how much presentation time will be available for each visual. In general a projected slide can hold your viewers' attention for about 30 seconds. Some of your slides will require only a few seconds of viewing. If you decide that a long discussion (longer than 30 seconds) is needed for a specific slide topic, then plan two or more similar slides to refocus the viewers' attention to the slide content.

D8k: Copy Slides

Your slide series may call for slides to be made from maps, pictures, diagrams, charts, or printed material. The single-lens reflex camera works best for this slide-making process and, when mounted on a copy-stand, is a simple means of obtaining excellent slides. Remember to get a release when copying from copyrighted materials. Suggestions for copying are as follows:

- Use a copy stand or tripod to steady your camera.
- Use photoflood lamps or lamps with reflectors - check for even lighting.
- Use a light meter.
- Use f/11 or f/16 to secure an adequate depth of field.

D8l: Title Slides

Title slides should introduce the viewer to the slide program and may also introduce subsections or emphasize a particular slide. Keep these slides brief and make certain that all type is legible. In preparing your lettering and artwork, keep in mind the correct proportion of your slides, use good design features, and select appropriate background colors.

D8m: Editing

Once the slides are prepared you are ready to begin editing. You may have taken some shots out of sequence, shot duplicate scenes, or have some omissions. The original script was a planning tool to give direction to the scene selection. Does it still fit? The editing step allows you to make changes, to refine your narration, to bring the visual and verbal content of the presentation into a final form.

Examine all your slides carefully. Make decisions about your slides and eliminate those that fail to contribute to your presentation or that are not of acceptable quality. If you change the planned presentation, re-edit the slides to fit the changes.

Keep in mind your anticipated audience as you edit your slide series. Let the pictures tell the greater part of your story. If you do not, you may only have a lecture with some illustrations and lose the maximum effect that good slides can provide your presentations.

Above all, place your slides in a projector carousel or cartridge and check the orientation in advance. There is no excuse for the upside-down or reversed slides caused by last minute loading.

D8n: <u>REMEMBER</u>

Visual aids can help you make an effective presentation. When you plan, organize, and develop your visuals, you are making decisions that enable you to be a more effective communicator. The time you spend in planning, producing, and arranging the visual dimension of your presentation reflects on your desire to be an effective communicator. Many viewers will make judgments about your professionalism based on how well you use audiovisuals in your presentation.

CHAPTER D9
REFERENCES FOR GRAPHIC COMMUNICATIONS

Adams, J. and D. Faux. Printing Technology - A Medium of Visual Communication. 2nd ed. Florence, KY: Brenton Publishers, 1982.

Besterfield, Dale and Robert O'Hagan. Technical Sketching. Reston, VA: Reston Publishing Co., 1983.

Brown, J.W., R.B. Lewis, and F.F. Harcleroad. AV Instruction: Technology, Media and Methods. 5th ed. New York: McGraw-Hill Co., 1977.

Brummitt, Wyatt. Photography Is... Garden City, NY: American Photographic Book Publishing Co., Inc., 1973.

French, Thomas E., and Charles J. Vierck. Graphic Science and Design. 4th ed. New York: McGraw Hill Co., 1984.

Hayes, Paul W. and Scott M. Worton. Essentials of Photography. Indianapolis, IN: Howard W. Sams & Co., Inc., 1983.

Kemp, Jerrold. Planning and Producing Audio Visual Materials. 3rd ed. New York: Thomas Y. Crovell, 1975.

Knowlton, Kenneth, and Robert Beauchemin. Technical Freehand Drawing. New York: McGraw-Hill Co., 1977.

Meilach, Dona Z. Dynamics of Presentation Graphics. Homewood, IL: Dow Jones-Irwin, 1986.

Minor, Ed and Harvey R. Frye. Techniques for Producing Visual Instructional Media. New York: McGraw-Hill Co., 1970.

Rowbotham, George (ed.). Engineering and Industrial Graphics Handbook. New York: McGraw-Hill Co., 1982.

Steidel, Robert F., and Jerald M. Henderson. The Graphic Languages of Engineering. New York: John Wiley & Sons, 1983.

Pamphlets:

Adventures in Color Slide Photography, Publication #AE-8. Rochester, NY: Eastman Kodak Co.

Adventures in Existing Light Photography, Publication #AC-44. Rochester, NY: Eastman Kodak Co.

Basic Developing, Printing and Enlarging, Publication #AJ-2. Rochester, NY: Eastman Kodak Co.

Basic Copying, Publication #AM-Z. Rochester, NY: Eastman Kodak Co.

Effective Lecture Slides, Publication #S-22. Rochester, NY:
Eastman Kodak Co.

Filters and Lens Attachments for Black and White and Color
Pictures, Publication #AB-1. Rochester, NY: Eastman Kodak
Co.

Indoor Picture Taking, Publication #AC-31. Rochester, NY:
Eastman Kodak Co.

Kodak Films, Color and Black and White, Publication #AF-1.
Rochester, NY: Eastman Kodak Co.

Making Black and White and Color Transparencies for Overhead
Projection, Publication #S-7. Rochester, NY: Eastman Kodak
Co.

Slides With a Purpose, Publication #Vi-15. Rochester, NY:
Eastman Kodak Co.

NOTES